Cram101 Textbook Outlines to accompany:

An Introduction to Brain and Behavior

Kolb and Whishaw, 1st Edition

An Academic Internet Publishers (AIPI) publication (c) 2007.

You have a discounted membership at www.Cram101.com with this book.

Get all of the practice tests for the chapters of this textbook, and access in-depth reference material for writing essays and papers. Here is an example from a Cram101 Biology text:

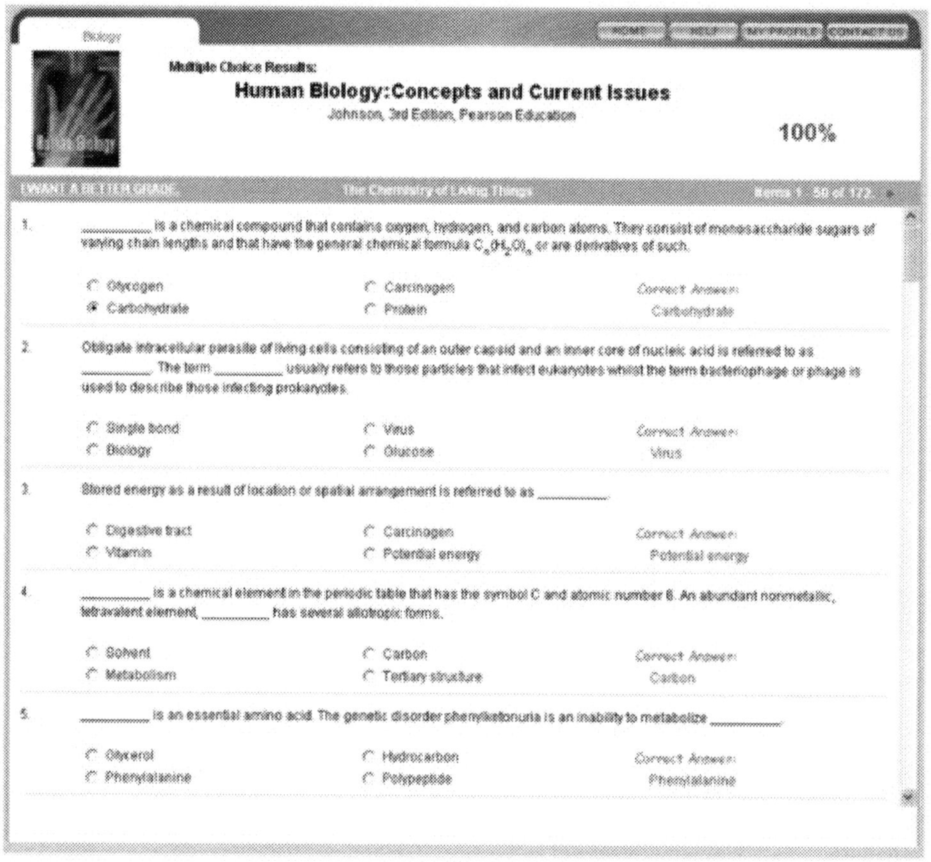

When you need problem solving help with math, stats, and other disciplines, www.Cram101.com will walk through the formulas and solutions step by step.

With Cram101.com online, you also have access to extensive reference material.

You will nail those essays and papers. Here is an example from a Cram101 Biology text:

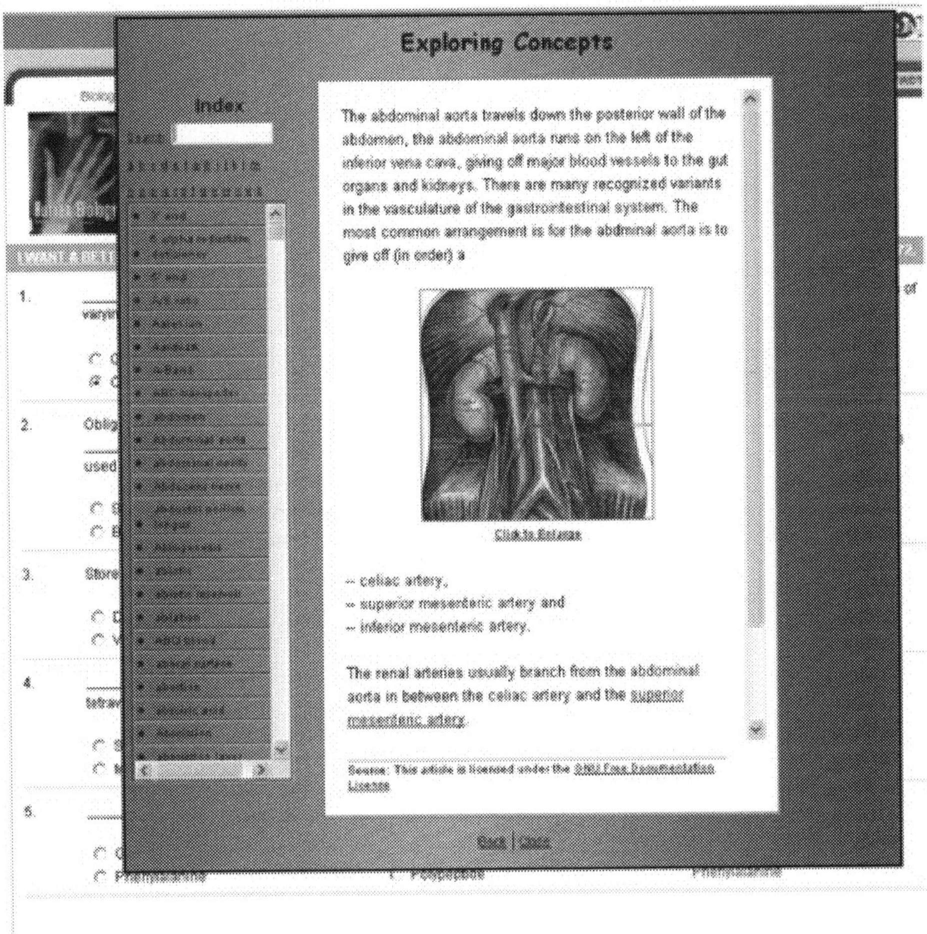

Visit **www.Cram101.com**, click Sign Up at the top of the screen, and enter DK73DW in the promo code box on the registration screen. Access to www.Cram101.com is normally $9.95, but because you have purchased this book, your access fee is only $4.95. Sign up and stop highlighting textbooks forever.

Learning System

Cram101 Textbook Outlines is a learning system. The notes in this book are the highlights of your textbook, you will never have to highlight a book again.

How to use this book. Take this book to class, it is your notebook for the lecture. The notes and highlights on the left hand side of the pages follow the outline and order of the textbook. All you have to do is follow along while your intructor presents the lecture. Circle the items emphasized in class and add other important information on the right side. With Cram101 Textbook Outlines you'll spend less time writing and more time listening. Learning becomes more efficient.

Cram101.com Online

Increase your studying efficiency by using Cram101.com's practice tests and online reference material. It is the perfect complement to Cram101 Textbook Outlines. Use self-teaching matching tests or simulate in-class testing with comprehensive multiple choice tests, or simply use Cram's true and false tests for quick review. Cram101.com even allows you to enter your in-class notes for an integrated studying format combining the textbook notes with your class notes.

Visit **www.Cram101.com**, click Sign Up at the top of the screen, and enter **DK73DW1919** in the promo code box on the registration screen. Access to www.Cram101.com is normally $9.95, but because you have purchased this book, your access fee is only $4.95. Sign up and stop highlighting textbooks forever.

 ISBN: 1-4288-0159-6

An Introduction to Brain and Behavior
Kolb and Whishaw, 1st

CONTENTS

Clinical psychologist	A psychologist, usually with a Ph.D, whose training is in the diagnosis, treatment, or research of psychological and behavioral disorders is a clinical psychologist.
Brain	The brain controls and coordinates most movement, behavior and homeostatic body functions such as heartbeat, blood pressure, fluid balance and body temperature. Functions of the brain are responsible for cognition, emotion, memory, motor learning and other sorts of learning. The brain is primarily made up of two types of cells: glia and neurons.
Consciousness	The awareness of the sensations, thoughts, and feelings being experienced at a given moment is called consciousness.
Affect	A subjective feeling or emotional tone often accompanied by bodily expressions noticeable to others is called affect.
Contusion	Brain contusion, a form of traumatic brain injury, is a bruise of the brain tissue. Like bruises in other tissues, cerebral contusion can be caused by multiple microhemorrhages, small blood vessel leaks into brain tissue.
Attention	Attention is the cognitive process of selectively concentrating on one thing while ignoring other things. Psychologists have labeled three types of attention: sustained attention, selective attention, and divided attention.
Evolution	Commonly used to refer to gradual change, evolution is the change in the frequency of alleles within a population from one generation to the next. This change may be caused by different mechanisms, including natural selection, genetic drift, or changes in population (gene flow).
Species	Species refers to a reproductively isolated breeding population.
Trait	An enduring personality characteristic that tends to lead to certain behaviors is called a trait. The term trait also means a genetically inherited feature of an organism.
Cerebral cortex	The cerebral cortex is the outermost layer of the cerebrum and has a grey color. It is made up of four lobes and it is involved in many complex brain functions including memory, perceptual awareness, "thinking", language and consciousness. The cerebral cortex receives sensory information from many different sensory organs eg: eyes, ears, etc. and processes the information.
Gyrus	A gyrus is a ridge on the cerebral cortex. It is generally surrounded by one or more sulci.
Lobes	The four major sections of the cerebral cortex: frontal, parietal, temporal, and occipital are called lobes.
Temporal lobe	The temporal lobe is part of the cerebrum. It lies at the side of the brain, beneath the lateral or Sylvian fissure. Adjacent areas in the superior, posterior and lateral parts of the temporal lobe are involved in high-level auditory processing.
Frontal lobe	The frontal lobe comprises four major folds of cortical tissue: the precentral gyrus, superior gyrus and the middle gyrus of the frontal gyri, the inferior frontal gyrus. It has been found to play a part in impulse control, judgement, language, memory, motor function, problem solving, sexual behavior, socialization and spontaneity.
Occipital lobe	The occipital lobe is the smallest of four true lobes in the human brain. Located in the rearmost portion of the skull, the occipital lobe is part of the forebrain structure. It is the visual processing center.
Parietal lobe	The parietal lobe is positioned above (superior to) the occipital lobe and behind (posterior to) the frontal lobe. It plays important roles in integrating sensory information from various senses, and in the manipulation of objects.
Central nervous system	The vertebrate central nervous system consists of the brain and spinal cord.

Left hemisphere	The left hemisphere of the cortex controls the right side of the body, coordinates complex movements, and, in 95% of people, controls the production of speech and written language.
Neuron	The neuron is the primary cell of the nervous system. They are found in the brain, the spinal cord, in the nerves and ganglia of the peripheral nervous system. It is a specialized cell that conducts impulses through the nervous system and contains three major parts: cell body, dendrites, and an axon. It can have many dendrites but only one axon.
Nerve	A nerve is an enclosed, cable-like bundle of nerve fibers or axons, which includes the glia that ensheath the axons in myelin. Neurons are sometimes called nerve cells, though this term is technically imprecise since many neurons do not form nerves.
Spinal cord	The spinal cord is a part of the vertebrate nervous system that is enclosed in and protected by the vertebral column (it passes through the spinal canal). It consists of nerve cells. The spinal cord carries sensory signals and motor innervation to most of the skeletal muscles in the body.
Peripheral nervous system	The peripheral nervous system consists of the nerves and neurons that serve the limbs and organs. It is not protected by bone or the blood-brain barrier, leaving it exposed to toxins and mechanical injuries. The peripheral nervous system is divided into the somatic nervous system and the autonomic nervous system.
Nervous system	The body's electrochemical communication circuitry, made up of billions of neurons is a nervous system.
Sensory nerves	Sensory nerves bring impulses toward the central nervous system.
Senses	The senses are systems that consist of a sensory cell type that respond to a specific kind of physical energy, and that correspond to a defined region within the brain where the signals are received and interpreted.
Sensory receptor	A sensory receptor is a structure that recognizes a stimulus in the environment of an organism. In response to stimuli the sensory receptor initiates sensory transduction by creating graded potentials or action potentials in the same cell or in an adjacent one.
Construct	A generalized concept, such as anxiety or gravity, is a construct.
Self-identity	The self-identity is the mental and conceptual awareness and persistent regard that sentient beings hold with regard to their own being.
Motor nerve	A motor nerve enables the brain to stimulate muscle contraction. A motor nerve is an efferent nerve that exclusively contains the axons of motorneurons, which innervate skeletal muscle.
Lungs	The lungs are the essential organs of respiration. Its principal function is to transport oxygen from the atmosphere into the bloodstream, and excrete carbon dioxide from the bloodstream into the atmosphere.
Autonomic nervous system	A division of the peripheral nervous system, the autonomic nervous system, regulates glands and activities such as heartbeat, respiration, digestion, and dilation of the pupils. It is responsible for homeostasis, maintaining a relatively constant internal environment .
Gland	A gland is an organ in an animal's body that synthesizes a substance for release such as hormones, often into the bloodstream or into cavities inside the body or its outer surface.
Larynx	The larynx, or voicebox, is an organ in the neck of mammals involved in protection of the trachea and sound production. The larynx houses the vocal cords, and is situated at the point where the upper tract splits into the trachea and the esophagus.
Sensory deprivation	Sensory deprivation is the deliberate reduction or removal of stimuli from one or more of the senses. Though short periods of sensory deprivation can be relaxing, extended deprivation can result in extreme anxiety, hallucinations, bizarre thoughts, depression, and antisocial behavior.
Hebb	Hebb demonstrated that the rearing of rats in an enriched environment could alter neural development

	and that sensory - neural connections were shaped by experience. He is famous for developing the concept of neural nets. He also believed that learning early in life is of the incremental variety, whereas later it is cognitive, insightful, and more all-or-none.
Social isolation	Social isolation refers to a type of loneliness that occurs when a person lacks a sense of integrated involvement. Being deprived of participation in a group or community involving companionship, shared interests, organized activities, and meaningful roles causes a person to feel alone.
Learning	Learning is a relatively permanent change in behavior that results from experience. Thus, to attribute a behavioral change to learning, the change must be relatively permanent and must result from experience.
Cones	Cones are photoreceptors that transmit sensations of color, function in bright light, and used in visual acutity. Infants prior to months of age can only distinguish green and red indicating the cones are not fully developed; they can see all of the colors by 2 months of age.
Innate	Innate behavior is not learned or influenced by the environment, rather, it is present or predisposed at birth.
Aristotle	Aristotle can be credited with the development of the first theory of learning. He concluded that ideas were generated in consciousness based on four principlesof association: contiguity, similarity, contrast, and succession. In contrast to Plato, he believed that knowledge derived from sensory experience and was not inherited.
Psyche	Psyche is the soul, spirit, or mind as distinguished from the body. In psychoanalytic theory, it is the totality of the id, ego, and superego, including both conscious and unconscious components.
Hypothesis	A specific statement about behavior or mental processes that is testable through research is a hypothesis.
Perception	Perception is the process of acquiring, interpreting, selecting, and organizing sensory information.
Emotion	An emotion is a mental states that arise spontaneously, rather than through conscious effort. They are often accompanied by physiological changes.
Scientific method	Psychologists gather data in order to describe, understand, predict, and control behavior. Scientific method refers to an approach that can be used to discover accurate information. It includes these steps: understand the problem, collect data, draw conclusions, and revise research conclusions.
Descartes	Descartes was concerned with the sharp contrast between the certainty of mathematics and the controversial nature of philosophy, and came to believe that the sciences could be made to yield results as certain as those of mathematics. He introduced the method of rationalism for arriving at knowledge. He also saw the human condition as a competition between the body and soul, introducing the concept of dualism.
Dualism	Dualism is a set of beliefs which begins with the claim that the mental and the physical have a fundamentally different nature. It is contrasted with varying kinds of monism, including materialism and phenomenalism. Dualism is one answer to the mind-body problem.
Neuroscience	A field that combines the work of psychologists, biologists, biochemists, medical researchers, and others in the study of the structure and function of the nervous system is neuroscience.
Lesion	A lesion is a non-specific term referring to abnormal tissue in the body. It can be caused by any disease process including trauma (physical, chemical, electrical), infection, neoplasm, metabolic and autoimmune.
Phineas Gage	As a result of an injury to his brain, Phineas Gage reportedly had significant changes in personality and temperament, which provided some of the first evidence that specific parts of the brain, particularly the frontal lobes, might be involved in specific psychological processes dealing with emotion, personality and problem solving.

Biological rhythm	A biological rhythm is a hypothetical cyclic pattern of alterations in physiology, emotions, and/or intellect
Reasoning	Reasoning is the act of using reason to derive a conclusion from certain premises. There are two main methods to reach a conclusion,deductive reasoning and inductive reasoning.
Harlow	Harlow and his famous wire and cloth surrogate mother monkey studies demonstrated that the need for affection created a stronger bond between mother and infant than did physical needs. He also found that the more discrimination problems the monkeys solved, the better they became at solving them.
Asylums	Asylums are hospitals specializing in the treatment of persons with mental illness. Psychiatric wards differ only in that they are a unit of a larger hospital.
Attitude	An enduring mental representation of a person, place, or thing that evokes an emotional response and related behavior is called attitude.
Sigmund Freud	Sigmund Freud was the founder of the psychoanalytic school, based on his theory that unconscious motives control much behavior, that particular kinds of unconscious thoughts and memories are the source of neurosis, and that neurosis could be treated through bringing these unconscious thoughts and memories to consciousness in psychoanalytic treatment.
Charcot	Charcot took an interest in the malady then called hysteria. It seemed to be a mental disorder with physical manifestations, of immediate interest to a neurologist. He believed that hysteria was the result of a weak neurological system which was hereditary.
Psychoanalysis	Psychoanalysis refers to the school of psychology that emphasizes the importance of unconscious motives and conflicts as determinants of human behavior. It was Freud's method of exploring human personality.
Pathology	Pathology is the study of the processes underlying disease and other forms of illness, harmful abnormality, or dysfunction.
Phenotype	The phenotype of an individual organism is either its total physical appearance and constitution, or a specific manifestation of a trait, such as size or eye color, that varies between individuals. Phenotype is determined to some extent by genotype, or by the identity of the alleles that an individual carries at one or more positions on the chromosomes.
Natural selection	Natural selection is a process by which biological populations are altered over time, as a result of the propagation of heritable traits that affect the capacity of individual organisms to survive and reproduce.
Allele	An allele is any one of a number of alternative forms of the same gene (sometimes the term refers to a non-gene sequence) occupying a given locus (position) on a chromosome.
Gene	A gene is an ultramicroscopic area of the chromosome. It is the smallest physical unit of the DNA molecule that carries a piece of hereditary information.
Darwin	Darwin achieved lasting fame as originator of the theory of evolution through natural selection. His book Expression of Emotions in Man and Animals is generally considered the first text on comparative psychology.
Adaptation	Adaptation is a lowering of sensitivity to a stimulus following prolonged exposure to that stimulus. Behavioral adaptations are special ways a particular organism behaves to survive in its natural habitat.
Population	Population refers to all members of a well-defined group of organisms, events, or things.
Mendel	Mendel is often called the "father of genetics" for his study of the inheritance of traits in pea plants. Mendel showed that there was particulate inheritance of traits according to his laws of inheritance.
Organelle	An organelle is one of several structures with specialized functions, suspended in the cytoplasm of a

	eukaryotic cell.
Simple cell	A neuron in the striate cortex that is maximally sensitive to the position and orientation of edges in the receptive field is called a simple cell.
Ganglion	A ganglion is a tissue mass that contains the dendrites and cell bodies (or "somas") of nerve cells, in most case ones belonging to the peripheral nervous system. Within the central nervous system such a mass is often called a nucleus.
Embryo	A developed zygote that has a rudimentary heart, brain, and other organs is referred to as an embryo.
Cerebellum	The cerebellum is located in the inferior posterior portion of the head (the hindbrain), directly dorsal to the brainstem and pons, inferior to the occipital lobe. The cerebellum is a region of the brain that plays an important role in the integration of sensory perception and fine motor output.
Relative size	Relative size is a monocular cue where an an automobile that is close to us seems larger than one that is far away. The visual system exploits the relative size of similar (or familiar) objects to judge distance.
Habit	A habit is a response that has become completely separated from its eliciting stimulus. Early learning theorists used the term to describe S-R associations, however not all S-R associations become a habit, rather many are extinguished after reinforcement is withdrawn.
Lewin	Lewin ranks as one of the pioneers of social psychology, as one of the founders of group dynamics and as one of the most eminent representatives of Gestalt psychology.
Elaboration	The extensiveness of processing at any given level of memory is called elaboration. The use of elaboration changes developmentally. Adolescents are more likely to use elaboration spontaneously than children.
Neuropsychologist	A psychologist concerned with the relationships among cognition, affect, behavior, and brain function is a neuropsychologist.
Glucose	Glucose, a simple monosaccharide sugar, is one of the most important carbohydrates and is used as a source of energy in animals and plants. Glucose is one of the main products of photosynthesis and starts respiration.
Maturation	The orderly unfolding of traits, as regulated by the genetic code is called maturation.
Stages	Stages represent relatively discrete periods of time in which functioning is qualitatively different from functioning at other periods.
Anatomy	Anatomy is the branch of biology that deals with the structure and organization of living things. It can be divided into animal anatomy (zootomy) and plant anatomy (phytonomy). Major branches of anatomy include comparative anatomy, histology, and human anatomy.
Neonatal period	The first 4 weeks of life, a time of transition from intrauterine dependency to independent existence is called the neonatal period.
Prenatal	Prenatal period refers to the time from conception to birth.
Stephen Jay Gould	Stephen Jay Gould helped Niles Eldredge develop the theory of punctuated equilibrium in 1972, wherein evolutionary change occurs relatively rapidly in comparatively brief periods of environmental stress, separated by longer periods of evolutionary stability.
Species-typical behavior	Species-typical behavior refers to any behavior pattern that is so characteristic of a given species of animal that it can be used to help identify that species.
Motivation	In psychology, motivation is the driving force (desire) behind all actions of an organism.
Correlation	A statistical technique for determining the degree of association between two or more variables is referred to as correlation.

Society	The social sciences use the term society to mean a group of people that form a semi-closed (or semi-open) social system, in which most interactions are with other individuals belonging to the group.
Dyslexia	Dyslexia is a neurological disorder with biochemical and genetic markers. In its most common and apparent form, it is a disability in which a person's reading and/or writing ability is significantly lower than that which would be predicted by his or her general level of intelligence.
Testosterone	Testosterone is a steroid hormone from the androgen group. It is the principal male sex hormone and the "original" anabolic steroid.
Hormone	A hormone is a chemical messenger from one cell (or group of cells) to another. The best known are those produced by endocrine glands, but they are produced by nearly every organ system. The function of hormones is to serve as a signal to the target cells; the action of the hormone is determined by the pattern of secretion and the signal transduction of the receiving tissue.
Acquisition	Acquisition is the process of adapting to the environment, learning or becoming conditioned. In classical conditoning terms, it is the initial learning of the stimulus response link, which involves a neutral stimulus being associated with a unconditioned stimulus and becoming a conditioned stimulus.
Stimulus	A change in an environmental condition that elicits a response is a stimulus.
Skeletal muscle	Skeletal muscle is a type of striated muscle, attached to the skeleton. They are used to facilitate movement, by applying force to bones and joints; via contraction. They generally contract voluntarily (via nerve stimulation), although they can contract involuntarily.
Individual differences	Individual differences psychology studies the ways in which individual people differ in their behavior. This is distinguished from other aspects of psychology in that although psychology is ostensibly a study of individuals, modern psychologists invariably study groups.
Sperry	Sperry separated the corpus callosum, the area of the brain used to transfer signals between the right and left hemispheres, to treat epileptics. He then tested these patients with tasks that were known to be dependant on specific hemispheres of the brain and demonstrated that the two halves of the brain now had independent functions.
Origin of Species	The Origin of Species by Charles Darwin makes the argument that groups of organisms gradually evolve through the process of natural selection. Characteristics that favor survival and reproduction are passed on to the next generation, those that do not, are gradually lost.
Instinct	Instinct is the word used to describe inherent dispositions towards particular actions. They are generally an inherited pattern of responses or reactions to certain kinds of situations.
Ethology	Where comparative psychology sees the study of animal behavior in the context of what is known about human psychology, ethology sees the study of animal behavior in the context of what is known about animal anatomy and physiology.
Lorenz	Lorenz demonstrated how incubator-hatched geese would imprint on the first suitable moving stimulus they saw within what he called a "critical period" of about 36 hours shortly after hatching. Most famously, the goslings would imprint on Lorenz himself .
Encephalitis	Encephalitis is an acute inflammation of the brain, commonly caused by a viral infection.
Meningitis	Meningitis is inflammation of the membranes covering the brain and the spinal cord. Although the most common causes are infection (bacterial, viral, fungal or parasitic), chemical agents and even tumor cells may cause meningitis.
Stroke	A stroke occurs when the blood supply to a part of the brain is suddenly interrupted by occlusion, by hemorrhage, or other causes

Cerebral cortex	The cerebral cortex is the outermost layer of the cerebrum and has a grey color. It is made up of four lobes and it is involved in many complex brain functions including memory, perceptual awareness, "thinking", language and consciousness. The cerebral cortex receives sensory information from many different sensory organs eg: eyes, ears, etc. and processes the information.
Brain	The brain controls and coordinates most movement, behavior and homeostatic body functions such as heartbeat, blood pressure, fluid balance and body temperature. Functions of the brain are responsible for cognition, emotion, memory, motor learning and other sorts of learning. The brain is primarily made up of two types of cells: glia and neurons.
Cerebellum	The cerebellum is located in the inferior posterior portion of the head (the hindbrain), directly dorsal to the brainstem and pons, inferior to the occipital lobe. The cerebellum is a region of the brain that plays an important role in the integration of sensory perception and fine motor output.
Species	Species refers to a reproductively isolated breeding population.
Olfactory bulb	The olfactory bulb is a part of the brain that is a distinct outgrowth from the forebrain. It plays a major role in olfaction, which is the perception of smells. The olfactory bulb receives direct input from olfactory nerve, made up of the axons from approximately 10 million olfactory receptor neurons in the olfactory mucosa, a region of the nasal cavity.
Perception	Perception is the process of acquiring, interpreting, selecting, and organizing sensory information.
Subjective experience	Subjective experience refers to reality as it is perceived and interpreted, not as it exists objectively.
Nervous system	The body's electrochemical communication circuitry, made up of billions of neurons is a nervous system.
Eardrum	The tympanum or tympanic membrane, colloquially known as the eardrum, is a thin membrane that separates the outer ear from the middle ear. Its function is to transmit sound from the air to the ossicles inside the middle ear. The malleus bone connects the eardrum to the other ossicles.
Evolution	Commonly used to refer to gradual change, evolution is the change in the frequency of alleles within a population from one generation to the next. This change may be caused by different mechanisms, including natural selection, genetic drift, or changes in population (gene flow).
Anatomy	Anatomy is the branch of biology that deals with the structure and organization of living things. It can be divided into animal anatomy (zootomy) and plant anatomy (phytonomy). Major branches of anatomy include comparative anatomy, histology, and human anatomy.
Efferent nerve	An efferent nerve carries nerve impulses away from the central nervous system. A motor nerve is an efferent nerve involved in muscular control.
Afferent nerve	Axons that carry information inward to the central nervous system from the periphery of the body are called afferent nerve fibers.
Gyrus	A gyrus is a ridge on the cerebral cortex. It is generally surrounded by one or more sulci.
Nucleus	In neuroanatomy, a cluster of cell bodies of neurons within the central nervous system is a nucleus.
Dura mater	The dura mater is the tough and inflexible outermost of the three layers of the meninges surrounding the brain. The other two meninges are the pia mater and the arachnoid mater. The dura mater envelops and protects the brain and spinal cord.
Central nervous	The vertebrate central nervous system consists of the brain and spinal cord.

system	
Sulcus	A sulcus is a depression or fissure in the surface of an organ, most especially the brain. In the brain it surrounds the gyri, creating the characteristic appearance of the brain.
Cerebrospinal fluid	A solution that fills the hollow cavities of the brain and circulates around the brain and spinal cord is called cerebrospinal fluid.
Meningitis	Meningitis is inflammation of the membranes covering the brain and the spinal cord. Although the most common causes are infection (bacterial, viral, fungal or parasitic), chemical agents and even tumor cells may cause meningitis.
Meninges	The meninges are the system of membranes that envelop the central nervous system. The meninges consist of three layers, the dura mater, the arachnoid mater, and the pia mater.
Cerebral hemisphere	Either of the two halves that make up the cerebrum is referred to as a cerebral hemisphere. The hemispheres operate together, linked by the corpus callosum, a very large bundle of nerve fibers, and also by other smaller commissures.
Hypothalamus	The hypothalamus is a region of the brain located below the thalamus, forming the major portion of the ventral region of the diencephalon and functioning to regulate certain metabolic processes and other autonomic activities.
Hindbrain	The lowest level of the brain, consisting of the medulla, cerebellum, and pons is called the hindbrain.
Midbrain	Located between the hindbrain and forebrain, a region in which many nerve-fiber systems ascend and descend to connect the higher and lower portions of the brain is referred to as midbrain. It is archipallian in origin, meaning its general architecture is shared with the most ancient of vertebrates. Dopamine produced in the subtantia nigra plays a role in motivation and habituation of species from humans to the most elementary animals such as insects.
Cerebrum	The cerebrum (the portion of the brain that performs motor and sensory functions and a variety of mental activities) is divided into four lobes - the frontal, temporal, parietal and occipital lobes.
Longitudinal fissure	The longitudinal fissure is the fissure (groove) that runs from the rostral to caudal portion of the brain, that serves to separate the left and right hemispheres
Lateral fissure	The lateral fissure is one of the most prominent structures of the human brain. It divides the frontal lobe and parietal lobe above from the temporal lobe below.
Central sulcus	The central sulcus is a prominent landmark of the brain, separating the parietal lobe from the frontal lobe. The central sulcus is the site of the primary motor cortex in mammals, a group of cells that controls voluntary movements of the body.
Lungs	The lungs are the essential organs of respiration. Its principal function is to transport oxygen from the atmosphere into the bloodstream, and excrete carbon dioxide from the bloodstream into the atmosphere.
Axon	An axon, or "nerve fiber," is a long slender projection of a nerve cell, or "neuron," which conducts electrical impulses away from the neuron's cell body or soma.
Lateral ventricles	Lateral ventricles are cavities in the brain filled with cerebrospinal fluid. Some individuals with schizophrenia have enlarged ventricles, probably resulting from insufficient development or atrophy of surrounding tissue.
Homogeneous	In biology homogeneous has a meaning similar to its meaning in mathematics. Generally it means "the same" or "of the same quality or general property".

White matter	White matter is one of the two main solid components of the central nervous system. It is composed of axons which connect various grey matter areas of the brain to each other and carry nerve impulses between neurons.
Capillary	The capillary is the smallest of a body's blood vessels, measuring 5-10 ìm. They connect arteries and veins, and most closely interact with tissues.
Corpus callosum	The corpus callosum is the largest white matter structure in the brain. It consists of mostly of contralateral axon projections. The corpus callosum connects the left and right cerebral hemispheres. Most communication between regions in different halves of the brain are carried over the corpus callosum.
Neocortex	The neocortex is part of the cerebral cortex which covers most of the surface of the cerebral hemispheres including the frontal, parietal, occipital, and temporal lobes. Often seen as the hallmark of human intelligence, the role of this structure in the brain appears to be involved in conscious thought, spatial reasoning, and sensory perception.
Encephalitis	Encephalitis is an acute inflammation of the brain, commonly caused by a viral infection.
Encephalitis lethargica	Encephalitis lethargica is an atypical form of encephalitis characterized by high fever, headache, double vision, delayed physical and mental response, and lethargy. In acute cases, patients may enter coma. Patients may also experience abnormal eye movements, upper body weakness, muscular pains, tremors, neck rigidity, and behavioral changes including psychosis.
Substantia nigra	The substantia nigra is a portion of the midbrain thought to be involved in certain aspects of movement and attention. Degeneration of cells in this region is the principle pathology that underlies Parkinson's disease.
Astrocyte	An astrocyte, also known as astroglia, is a characteristic star-shaped cell in the brain. They are the biggest cells found in brain tissue and outnumber the neurons ten to one. A commonly accepted function is to structure physically the brain. A second function is to provide neurons with nutrients such as glucose. They regulate the flow of nutrients provided by capillaries by forming the blood-brain barrier.
Neuron	The neuron is the primary cell of the nervous system. They are found in the brain, the spinal cord, in the nerves and ganglia of the peripheral nervous system. It is a specialized cell that conducts impulses through the nervous system and contains three major parts: cell body, dendrites, and an axon. It can have many dendrites but only one axon.
Glial	Glial cells are non-neuronal cells that provide support and nutrition, maintain homeostasis, form myelin, and participate in signal transmission in the nervous system.
Nerve	A nerve is an enclosed, cable-like bundle of nerve fibers or axons, which includes the glia that ensheath the axons in myelin. Neurons are sometimes called nerve cells, though this term is technically imprecise since many neurons do not form nerves.
Spinal cord	The spinal cord is a part of the vertebrate nervous system that is enclosed in and protected by the vertebral column (it passes through the spinal canal). It consists of nerve cells. The spinal cord carries sensory signals and motor innervation to most of the skeletal muscles in the body.
Stroke	A stroke occurs when the blood supply to a part of the brain is suddenly interrupted by occlusion, by hemorrhage, or other causes
Prognosis	A forecast about the probable course of an illess is referred to as prognosis.
Depression	In everyday language depression refers to any downturn in mood, which may be relatively transitory and perhaps due to something trivial. This is differentiated from Clinical depression which is marked by symptoms that last two weeks or more and are so severe that they interfere with daily living.

Anxiety	Anxiety is a complex combination of the feeling of fear, apprehension and worry often accompanied by physical sensations such as palpitations, chest pain and/or shortness of breath.
Placebo	Placebo refers to a bogus treatment that has the appearance of being genuine.
Inference	Inference is the act or process of drawing a conclusion based solely on what one already knows.
Peripheral nervous system	The peripheral nervous system consists of the nerves and neurons that serve the limbs and organs. It is not protected by bone or the blood-brain barrier, leaving it exposed to toxins and mechanical injuries. The peripheral nervous system is divided into the somatic nervous system and the autonomic nervous system.
Functional analysis	A systematic study of behavior in which one identifies the stimuli that trigger the behavior and the reinforcers that maintain it is a functional analysis. Relations between the two become the cause-and-effect relationships in behavior and are the laws of a science. A synthesis of these various laws expressed in quantitative terms yields a comprehensive picture of the organism as a behaving system without postulating internal processes.
Learning	Learning is a relatively permanent change in behavior that results from experience. Thus, to attribute a behavioral change to learning, the change must be relatively permanent and must result from experience.
Reticular formation	Reticular formation is a part of the brain which is involved in stereotypical actions, such as walking, sleeping, and lying down. The reticular formation, phylogenetically one of the oldest portions of the brain, is a poorly-differentiated area of the brain stem.
Medulla oblongata	The medulla oblongata is the lower portion of the brainstem. It is above the spinal cord, below the pons, and anterior to the cerebellum. It controls autonomic functions and relays nerve signals between the brain and spinal cord.
Cranial nerves	Cranial nerves are nerves that emerge from the brainstem instead of the spinal cord. Cranial nerves I and II are named as such, but are technically not nerves, as they are continuations of the central nervous system.
Basal ganglia	The basal ganglia are a group of nuclei in the brain associated with motor and learning functions.
Cranial nerve	A Cranial nerve emerges from the brainstem instead of the spinal cord.
Limbic system	The limbic system is a group of brain structures that are involved in various emotions such as aggression, fear, pleasure and also in the formation of memory. The limbic system affects the endocrine system and the autonomic nervous system. It consists of several subcortical structures located around the thalamus.
Pons	The pons is a knob on the brain stem. It is part of the autonomic nervous system, and relays sensory information between the cerebellum and cerebrum. Some theories posit that it has a role in dreaming.
Epithalamus	The epithalamus is a dorsal posterior segment of the diencephalon. Some functions of its components include the secretion of melatonin by the pineal gland, and the regulation of hunger and thirst by the habenula.
Thalamus	An area near the center of the brain involved in the relay of sensory information to the cortex and in the functions of sleep and attention is the thalamus.
Forebrain	The forebrain is the highest level of the brain. Key structures in the forebrain are the limbic system, thalamus, basal ganglia, hypothalamus, and cerebral cortex.
Senses	The senses are systems that consist of a sensory cell type that respond to a specific kind of

	physical energy, and that correspond to a defined region within the brain where the signals are received and interpreted.
Diencephalon	The diencephalon is the region of the brain that includes the epithalamus, thalamus, and hypothalamus. It is located above the mesencephalon of the brain stem. Sensory information is relayed between the brain stem and the rest of the brain regions
Sensation	Sensation is the first stage in the chain of biochemical and neurologic events that begins with the impinging of a stimulus upon the receptor cells of a sensory organ, which then leads to perception, the mental state that is reflected in statements like "I see a uniformly blue wall."
Viscera	The internal organs of the body are called viscera.
Larynx	The larynx, or voicebox, is an organ in the neck of mammals involved in protection of the trachea and sound production. The larynx houses the vocal cords, and is situated at the point where the upper tract splits into the trachea and the esophagus.
Relative size	Relative size is a monocular cue where an an automobile that is close to us seems larger than one that is far away. The visual system exploits the relative size of similar (or familiar) objects to judge distance.
Reticular activating system	The reticular activating system is the part of the brain believed to be the center of arousal and motivation. It is situated between the brain stem and the central nervous system (CNS).
Gray matter	Gray matter is a category of nervous tissue with many nerve cell bodies and few myelinated axons. Generally, gray matter can be understood as the parts of the brain responsible for information processing; whereas, white matter is responsible for information transmission. In addition, gray matter does not have a myelin sheath and does not regenerate after injury unlike white matter.
Cardiovascular system	The human cardiovascular system comprises the blood, the heart, and a dual-circuit system of blood vessels that serve as conduits between the heart, the lungs, and the peripheral tissues of the body.
Optic nerve	The optic nerve is the nerve that transmits visual information from the retina to the brain. The optic nerve is composed of retinal ganglion cell axons and support cells.
Tactile	Pertaining to the sense of touch is referred to as tactile.
Pituitary gland	The pituitary gland is an endocrine gland about the size of a pea that sits in the small, bony cavity at the base of the brain. The pituitary gland secretes hormones regulating a wide variety of bodily activities, including trophic hormones that stimulate other endocrine glands.
Hormone	A hormone is a chemical messenger from one cell (or group of cells) to another. The best known are those produced by endocrine glands, but they are produced by nearly every organ system. The function of hormones is to serve as a signal to the target cells; the action of the hormone is determined by the pattern of secretion and the signal transduction of the receiving tissue.
Optic chiasm	Optic chiasm refers to the point at which the optic nerves from the inside half of each eye cross over and then project to the opposite half of the brain.
Lateral geniculate nucleus	The lateral geniculate nucleus of the thalamus is a part of the brain, which is the primary processor of visual information, received from the retina, in the CNS.
Frontal lobe	The frontal lobe comprises four major folds of cortical tissue: the precentral gyrus,

	superior gyrus and the middle gyrus of the frontal gyri, the inferior frontal gyrus. It has been found to play a part in impulse control, judgement, language, memory, motor function, problem solving, sexual behavior, socialization and spontaneity.
Analogy	An analogy is a comparison between two different things, in order to highlight some form of similarity. Analogy is the cognitive process of transferring information from a particular subject to another particular subject.
Cingulate cortex	The part of the limbic system that is believed to process cognitive information in emotion is the cingulate cortex. The cingulate cortex is part of the brain and situated roughly in the middle of the cortex.
Hippocampus	The hippocampus is a part of the brain located inside the temporal lobe. It forms a part of the limbic system and plays a part in memory and navigation.
Affective	Affective is the way people react emotionally, their ability to feel another living thing's pain or joy.
Lobes	The four major sections of the cerebral cortex: frontal, parietal, temporal, and occipital are called lobes.
Temporal lobe	The temporal lobe is part of the cerebrum. It lies at the side of the brain, beneath the lateral or Sylvian fissure. Adjacent areas in the superior, posterior and lateral parts of the temporal lobe are involved in high-level auditory processing.
Occipital lobe	The occipital lobe is the smallest of four true lobes in the human brain. Located in the rearmost portion of the skull, the occipital lobe is part of the forebrain structure. It is the visual processing center.
Tremor	Tremor is the rhythmic, oscillating shaking movement of the whole body or just a certain part of it, caused by problems of the neurons responsible from muscle action.
Motor cortex	Motor cortex refers to the section of cortex that lies in the frontal lobe, just across the central fissure from the sensory cortex. Neural impulses in the motor cortex are linked to muscular responses throughout the body.
Syndrome	The term syndrome is the association of several clinically recognizable features, signs, symptoms, phenomena or characteristics which often occur together, so that the presence of one feature indicates the presence of the others.
Tics	Tics are a repeated, impulsive action, almost reflexive in nature, which the person feels powerless to control or avoid.
Sigmund Freud	Sigmund Freud was the founder of the psychoanalytic school, based on his theory that unconscious motives control much behavior, that particular kinds of unconscious thoughts and memories are the source of neurosis, and that neurosis could be treated through bringing these unconscious thoughts and memories to consciousness in psychoanalytic treatment.
Theories	Theories are logically self-consistent models or frameworks describing the behavior of a certain natural or social phenomenon. They are broad explanations and predictions concerning phenomena of interest.
Emotion	An emotion is a mental states that arise spontaneously, rather than through conscious effort. They are often accompanied by physiological changes.
Hypothesis	A specific statement about behavior or mental processes that is testable through research is a hypothesis.
Cingulate gyrus	Cingulate gyrus is a gyrus in the medial part of the brain. It partially wraps around the corpus callosum and is limited above by the cingulate sulcus. It functions as an intergral part of the limbic system, which is involved with emotion formation and processing, learning,

	and memory.
Amygdala	Located in the brain's medial temporal lobe, the almond-shaped amygdala is believed to play a key role in the emotions. It forms part of the limbic system and is linked to both fear responses and pleasure. Its size is positively correlated with aggressive behavior across species.
Bedlam	Bedlam is the world's oldest psychiatric hospital. It became famous and afterwards infamous for the brutal ill-treatment meted out to the insane.
Olfactory system	The olfactory system is the sensory system used for the sense of smell. The olfactory system is often spoken of along with the gustatory system as the chemosensory senses because both transduce chemical signals into perception
Sensory receptor	A sensory receptor is a structure that recognizes a stimulus in the environment of an organism. In response to stimuli the sensory receptor initiates sensory transduction by creating graded potentials or action potentials in the same cell or in an adjacent one.
Nasal cavity	The nasal cavity is a large air-filled space above and behind the nose in the middle of the face. The nasal cavity is important in warming and cleaning the air as it is inhaled. The nasal cavity also contains organs involved in olfaction.
Receptor	A sensory receptor is a structure that recognizes a stimulus in the internal or external environment of an organism. In response to stimuli the sensory receptor initiates sensory transduction by creating graded potentials or action potentials in the same cell or in an adjacent one.
Sympathetic	The sympathetic nervous system activates what is often termed the "fight or flight response". It is an automatic regulation system, that is, one that operates without the intervention of conscious thought.
Neurologist	A physician who studies the nervous system, especially its structure, functions, and abnormalities is referred to as neurologist.
Glucose	Glucose, a simple monosaccharide sugar, is one of the most important carbohydrates and is used as a source of energy in animals and plants. Glucose is one of the main products of photosynthesis and starts respiration.
Liver	The liver plays a major role in metabolism and has a number of functions in the body including detoxification, glycogen storage and plasma protein synthesis. It also produces bile, which is important for digestion. The liver converts most carbohydrates, proteing, and fats into glucose.
Pupil	In the eye, the pupil is the opening in the middle of the iris. It appears black because most of the light entering it is absorbed by the tissues inside the eye. The size of the pupil is controlled by involuntary contraction and dilation of the iris, in order to regulate the intensity of light entering the eye. This is known as the pupillary reflex.
Gland	A gland is an organ in an animal's body that synthesizes a substance for release such as hormones, often into the bloodstream or into cavities inside the body or its outer surface.
Parasympathetic	The parasympathetic nervous system is one of two divisions of the autonomic nervous system. It conserves energy as it slows the heart rate, increases intestinal and gland activity, and relaxes sphincter muscles. In another words, it acts to reverse the effects of the Sympathetic nervous system.
Charles Bell	Charles Bell in 1811 discovered that the sensory fibers of a mized nerve enter the spinal cord at the dorsal or posterior nerve root, whereaas the motor fibers of the same nerve leave the spinal chord by a ventral or anterior root.

Facial nerve	The facial nerve is seventh of twelve paired cranial nerves. Its main function is motor control of most of the facial muscles and muscles of the inner ear.
Ganglion	A ganglion is a tissue mass that contains the dendrites and cell bodies (or "somas") of nerve cells, in most case ones belonging to the peripheral nervous system. Within the central nervous system such a mass is often called a nucleus.
Synapse	A synapse is specialized junction through which cells of the nervous system signal to one another and to non-neuronal cells such as muscles or glands.
Consciousness	The awareness of the sensations, thoughts, and feelings being experienced at a given moment is called consciousness.
Hue	A hue refers to the gradation of color within the optical spectrum, or visible spectrum, of light. Hue may also refer to a particular color within this spectrum, as defined by its dominant wavelength, or the central tendency of its combined wavelengths.
Right hemisphere	The brain is divided into left and right cerebral hemispheres. The right hemisphere of the cortex controls the left side of the body.
Schematic representation	The representation of objects in terms of real or potential interactions with other objects is called a schematic representation.
Somatosensory	Somatosensory system consists of the various sensory receptors that trigger the experiences labelled as touch or pressure, temperature, pain, and the sensations of muscle movement and joint position including posture, movement, and facial expression.
Projection	Attributing one's own undesirable thoughts, impulses, traits, or behaviors to others is referred to as projection.
Fixation	Fixation in abnormal psychology is the state where an individual becomes obsessed with an attachment to another human, animal or inanimate object. Fixation in vision refers to maintaining the gaze in a constant direction. .
Left hemisphere	The left hemisphere of the cortex controls the right side of the body, coordinates complex movements, and, in 95% of people, controls the production of speech and written language.
Elaboration	The extensiveness of processing at any given level of memory is called elaboration. The use of elaboration changes developmentally. Adolescents are more likely to use elaboration spontaneously than children.
American Sign Language	The American Sign Language is a language of hand gestures used by deaf and hearing impaired persons.
Scheme	According to Piaget, a hypothetical mental structure that permits the classification and organization of new information is called a scheme.
Autonomic nervous system	A division of the peripheral nervous system, the autonomic nervous system, regulates glands and activities such as heartbeat, respiration, digestion, and dilation of the pupils. It is responsible for homeostasis, maintaining a relatively constant internal environment .
Skeletal muscle	Skeletal muscle is a type of striated muscle, attached to the skeleton. They are used to facilitate movement, by applying force to bones and joints; via contraction. They generally contract voluntarily (via nerve stimulation), although they can contract involuntarily.
Brain circuits	Neurotransmitter currents or neural pathways in the brain are referred to as brain circuits.
Research method	The scope of the research method is to produce some new knowledge. This, in principle, can take three main forms: Exploratory research; Constructive research; and Empirical research.

Nervous system — The body's electrochemical communication circuitry, made up of billions of neurons is a nervous system.

Neuron — The neuron is the primary cell of the nervous system. They are found in the brain, the spinal cord, in the nerves and ganglia of the peripheral nervous system. It is a specialized cell that conducts impulses through the nervous system and contains three major parts: cell body, dendrites, and an axon. It can have many dendrites but only one axon.

Insight — Insight refers to a sudden awareness of the relationships among various elements that had previously appeared to be independent of one another.

Organelle — An organelle is one of several structures with specialized functions, suspended in the cytoplasm of a eukaryotic cell.

Learning — Learning is a relatively permanent change in behavior that results from experience. Thus, to attribute a behavioral change to learning, the change must be relatively permanent and must result from experience.

Brain — The brain controls and coordinates most movement, behavior and homeostatic body functions such as heartbeat, blood pressure, fluid balance and body temperature. Functions of the brain are responsible for cognition, emotion, memory, motor learning and other sorts of learning. The brain is primarily made up of two types of cells: glia and neurons.

Chromosome — The DNA which carries genetic information in biological cells is normally packaged in the form of one or more large macromolecules called a chromosome. Humans normally have 46.

Protein — A protein is a complex, high-molecular-weight organic compound that consists of amino acids joined by peptide bonds. It is essential to the structure and function of all living cells and viruses. Many are enzymes or subunits of enzymes.

Gene — A gene is an ultramicroscopic area of the chromosome. It is the smallest physical unit of the DNA molecule that carries a piece of hereditary information.

Research method — The scope of the research method is to produce some new knowledge. This, in principle, can take three main forms: Exploratory research; Constructive research; and Empirical research.

Cajal — Cajal, Nobel laureate, is considered to be the father of modern neuroscience. He postulated that the nervous system is made up of billions of separate nerve cells (neurons) and that nerve cells are polarized. He suggested that neurons communicate with each other via specialized junctions. This hypothesis became the basis of the neuron doctrine.

Golgi — Golgi discovered a method of staining nervous tissue which would stain a limited number of cells at random, in their entirety. This enabled him to view the paths of nerve cells in the brain for the first time. He called his discovery the black reaction. It is now known universally as the Golgi stain.

Staining — Staining is a biochemical technique of adding a class-specific (DNA, proteins, lipids, carbohydrates) dye to a substrate to qualify or quantify the presence of a specific compound. They are frequently used to highlight structures in tissues for viewing, often with the aid of different microscopes.

Nerve — A nerve is an enclosed, cable-like bundle of nerve fibers or axons, which includes the glia that ensheath the axons in myelin. Neurons are sometimes called nerve cells, though this term is technically imprecise since many neurons do not form nerves.

Hypothesis — A specific statement about behavior or mental processes that is testable through research is a hypothesis.

Dendrite — A dendrite is a slender, typically branched projection of a nerve cell, or "neuron," which conducts the electrical stimulation received from other cells to the body or soma of the cell

	from which it projects. This stimulation arrives through synapses, which typically are located near the tips of the dendrites and away from the soma.
Axon	An axon, or "nerve fiber," is a long slender projection of a nerve cell, or "neuron," which conducts electrical impulses away from the neuron's cell body or soma.
Glial	Glial cells are non-neuronal cells that provide support and nutrition, maintain homeostasis, form myelin, and participate in signal transmission in the nervous system.
Cell membrane	A component of every biological cell, the selectively permeable cell membrane is a thin and structured bilayer of phospholipid and protein molecules that envelopes the cell. It separates a cell's interior from its surroundings and controls what moves in and out.
Axon hillock	The axon hillock is the anatomical part of a neuron that connects the cell body to the axon. It is regarded as the place where EPSPs from numerous synaptic inputs on the dendrites or cell body accumulate.
Dendritic spine	A dendritic spine is a small membranous extrusion that protrudes from a dendrite and forms one half of a synapse. Changes in dendritic spine density underlie many brain functions, including motivation, learning, and memory. In particular, long-term memory is mediated in part by the growth of new dendritic spines to reinforce a particular neural pathway.
Synapse	A synapse is specialized junction through which cells of the nervous system signal to one another and to non-neuronal cells such as muscles or glands.
Soma	The soma, or cell body, is the bulbous end of a neuron, containing the nucleus. The cell nucleus is a key feature of the soma. The nucleus is the source of most of the RNA that is produced in neurons and most proteins are produced from mRNAs that do not travel far from the nucleus.
Nucleus	In neuroanatomy, a cluster of cell bodies of neurons within the central nervous system is a nucleus.
Purkinje cells	Purkinje cells are a class of GABAergic neuron located in the cerebellar cortex. Purkinje cells send inhibitory projections to the deep cerebellar nuclei, and constitute the sole output of all motor coordination in the cerebellar cortex.
Interneuron	An interneuron (also called relay neuron or association neuron) is a neuron that communicates only to other neurons.
Purkinje	Purkinje is best known for his 1837 discovery of Purkinje cells, large neurons with many branching fibres found in the cerebellum. He is also known for his discovery, in 1839 of Purkinje fibres, the fibrous tissue that conducts electrical impulses from the atrioventricular node to all parts of the ventricles of the heart.
Sensory receptor	A sensory receptor is a structure that recognizes a stimulus in the environment of an organism. In response to stimuli the sensory receptor initiates sensory transduction by creating graded potentials or action potentials in the same cell or in an adjacent one.
Sensory neuron	A sensory neuron is an afferent nerve cell within the nervous system responsible for converting external stimuli from the organism's environment into internal electrical impulses. It carries messages from a sensory organ, through a nerve, into the brain or spinal cord.
Motor neuron	A motor neuron is an efferent neuron that originates in the spinal cord and synapses with muscle fibers to facilitate muscle contraction and with muscle spindles to modify proprioceptive sensitivity.
Bipolar cell	As a part of the retina, the bipolar cell exists between photoreceptors and ganglion cells. Bipolar cells are so-named as they have a central body from which two sets of processes

arise. At one end, they form synapses with either a single cone cell, or a number of rod cells. At the other end, they form synapses with ganglion cells, which fire action potentials along the optic nerve (cranial nerve II). They effectively transfer information from rods and cones to ganglion cells.

Retina

The retina is a thin layer of cells at the back of the eyeball. It is the part of the eye which converts light into nervous signals. The retina contains photoreceptor cells which receive the light; the resulting neural signals then undergo complex processing by other neurons of the retina, and are transformed into action potentials in retinal ganglion cells whose axons form the optic nerve.

Photoreceptor

Photoreceptor cells are contained in the retina and are responsible for transducing, or converting, light into signals that can be ultimately transmitted to the brain via the optic nerve. A photoreceptor cell contains a membranous protein called an opsin which contains a pigment molecule called retinal.

Neocortex

The neocortex is part of the cerebral cortex which covers most of the surface of the cerebral hemispheres including the frontal, parietal, occipital, and temporal lobes. Often seen as the hallmark of human intelligence, the role of this structure in the brain appears to be involved in conscious thought, spatial reasoning, and sensory perception.

Cerebellum

The cerebellum is located in the inferior posterior portion of the head (the hindbrain), directly dorsal to the brainstem and pons, inferior to the occipital lobe. The cerebellum is a region of the brain that plays an important role in the integration of sensory perception and fine motor output.

Spinal cord

The spinal cord is a part of the vertebrate nervous system that is enclosed in and protected by the vertebral column (it passes through the spinal canal). It consists of nerve cells. The spinal cord carries sensory signals and motor innervation to most of the skeletal muscles in the body.

Feedback

Feedback refers to information returned to a person about the effects a response has had.

Excitatory message

An excitatory message is a chemical message that makes it more likely that a receiving neuron will fire and an action potential will travel down its axons.

Cerebrospinal fluid

A solution that fills the hollow cavities of the brain and circulates around the brain and spinal cord is called cerebrospinal fluid.

Neuropsychology

Neuropsychology is a branch of psychology that aims to understand how the structure and function of the brain relates to specific psychological processes.

Seizure

A seizure is a temporary alteration in brain function expressed as a changed mental state, tonic or clonic movements and various other symptoms. They are due to temporary abnormal electrical activity of a group of brain cells.

Frontal lobe

The frontal lobe comprises four major folds of cortical tissue: the precentral gyrus, superior gyrus and the middle gyrus of the frontal gyri, the inferior frontal gyrus. It has been found to play a part in impulse control, judgement, language, memory, motor function, problem solving, sexual behavior, socialization and spontaneity.

Tumor

A tumor is an abnormal growth that when located in the brain can either be malignant and directly destroy brain tissue, or be benign and disrupt functioning by increasing intracranial pressure.

Astrocyte

An astrocyte, also known as astroglia, is a characteristic star-shaped cell in the brain. They are the biggest cells found in brain tissue and outnumber the neurons ten to one. A commonly accepted function is to structure physically the brain. A second function is to provide neurons with nutrients such as glucose. They regulate the flow of nutrients provided

	by capillaries by forming the blood-brain barrier.
Chemotherapy	Chemotherapy is the use of chemical substances to treat disease. In its modern-day use, it refers almost exclusively to cytostatic drugs used to treat cancer.In its non-oncological use, the term may also refer to antibiotics.
Central nervous system	The vertebrate central nervous system consists of the brain and spinal cord.
Blood-brain barrier	The blood-brain barrier is a physical barrier between the blood vessels in the central nervous system, and most parts of the central nervous system itself. The blood-brain barrier appears to exist primarily to protect the brain from the chemical messenger systems flowing around the body.
Glucose	Glucose, a simple monosaccharide sugar, is one of the most important carbohydrates and is used as a source of energy in animals and plants. Glucose is one of the main products of photosynthesis and starts respiration.
Microglia	Microglia act as the immune cells of the Central nervous system (CNS).
Peripheral nervous system	The peripheral nervous system consists of the nerves and neurons that serve the limbs and organs. It is not protected by bone or the blood-brain barrier, leaving it exposed to toxins and mechanical injuries. The peripheral nervous system is divided into the somatic nervous system and the autonomic nervous system.
Sensation	Sensation is the first stage in the chain of biochemical and neurologic events that begins with the impinging of a stimulus upon the receptor cells of a sensory organ, which then leads to perception, the mental state that is reflected in statements like "I see a uniformly blue wall."
Schwann cell	A Schwann cell is a variety of neuroglia that wraps around axons in the peripheral nervous system, forming the myelin sheath. The nervous system depends crucially on this sheath for insulation and an increase in impulse speed.
Myelin	Myelin is an electrically insulating fatty layer that surrounds the axons of many neurons, especially those in the peripheral nervous system. The main consequence of a myelin sheath is an increase in the speed at which impulses propagate along the myelinated fiber. The sheath continues to develop throughout childhood.
Myelination	The process in which the nerve cells are covered and insulated with a layer of fat cells, which increases the speed at which information travels through the nervous system is referred to as myelination.
Multiple sclerosis	Multiple sclerosis affects neurons, the cells of the brain and spinal cord that carry information, create thought and perception, and allow the brain to control the body. Surrounding and protecting these neurons is a layer of fat, called myelin, which helps neurons carry electrical signals. MS causes gradual destruction of myelin (demyelination) in patches throughout the brain and/or spinal cord, causing various symptoms depending upon which signals are interrupted.
Neurologist	A physician who studies the nervous system, especially its structure, functions, and abnormalities is referred to as neurologist.
Local anesthetic	Local anesthetic drugs act mainly by inhibiting sodium influx through sodium-specific ion channels in the neuronal cell membrane, in particular the so-called voltage-gated sodium channels. When the influx of sodium is interrupted, an action potential cannot arise and signal conduction is thus inhibited.
Depression	In everyday language depression refers to any downturn in mood, which may be relatively transitory and perhaps due to something trivial. This is differentiated from Clinical

	depression which is marked by symptoms that last two weeks or more and are so severe that they interfere with daily living.
Affect	A subjective feeling or emotional tone often accompanied by bodily expressions noticeable to others is called affect.
Ion	An ion is an atom or group of atoms with a net electric charge. The energy required to detach an electron in its lowest energy state from an atom or molecule of a gas with less net electric charge is called the ionization potential, or ionization energy.
Enzyme	An enzyme is a protein that catalyzes, or speeds up, a chemical reaction. Enzymes are essential to sustain life because most chemical reactions in biological cells would occur too slowly, or would lead to different products, without enzymes.
Mitochondria	Mitochondria are sometimes described as "cellular power plants", because their primary function is to convert organic materials into energy in the form of ATP.
Motility	Motility is the ability to move spontaneously and independently. The term can apply to single cells, or to multicellular organisms.
Nucleotide	A nucleotide is a monomer or the structural unit of nucleotide chains forming nucleic acids as RNA and DNA. They also play important roles in cellular energy transport and transformations and in enzyme regulation.
Analogy	An analogy is a comparison between two different things, in order to highlight some form of similarity. Analogy is the cognitive process of transferring information from a particular subject to another particular subject.
Amino acid	Amino acid is the basic structural building unit of proteins. They form short polymer chains called peptides or polypeptides which in turn form structures called proteins.
Genetic code	The genetic code is a set of rules, which maps DNA sequences to proteins in the living cell, and is employed in the process of protein synthesis. Nearly all living things use the same genetic code, called the standard genetic code, although a few organisms use minor variations of the standard code.
Ribosome	A ribosome is an organelle composed of rRNA and ribosomal proteins. It translates mRNA into a polypeptide chain (e.g., a protein). It can be thought of as a factory that builds a protein from a set of genetic instructions.
Peptides	Peptides are the family of molecules formed from the linking, in a defined order, of various amino acids.
Receptor	A sensory receptor is a structure that recognizes a stimulus in the internal or external environment of an organism. In response to stimuli the sensory receptor initiates sensory transduction by creating graded potentials or action potentials in the same cell or in an adjacent one.
Receptor site	A location on the dendrite of a receiving neuron that is tailored to receive a specific neurotransmitter is a receptor site.
Mutation	Mutation is a permanent, sometimes transmissible (if the change is to a germ cell) change to the genetic material (usually DNA or RNA) of a cell. They can be caused by copying errors in the genetic material during cell division and by exposure to radiation, chemicals, or viruses, or can occur deliberately under cellular control during the processes such as meiosis or hypermutation.
Genetic disorder	A genetic disorder is a disease caused by abnormal expression of one or more genes in a person causing a clinical phenotype.
Autosome	An autosome is a non-sex chromosome. It is an ordinary paired chromosome that is the same in

both sexes of a species.

Sex chromosomes The sex chromosomes are the 23rd pair of chromosomes. They determine whether the child will be male or female. A pair with two X-shaped chromosomes produces a female. A pair with an X chromosome and a Y chromosome produces a male.

X chromosome The sex chromosomes are one of the 23 pairs of human chromosomes. Each person normally has one pair of sex chromosomes in each cell. Females have two X chromosomes, while males have one X and one Y chromosome. The X chromosome carries hundreds of genes but few, if any, of these have anything to do directly with sex determination.

Allele An allele is any one of a number of alternative forms of the same gene (sometimes the term refers to a non-gene sequence) occupying a given locus (position) on a chromosome.

Heterozygous Heterozygous refers to the condition in which a pair of genes occupying the same locus on a pair of chromosomes are different from one another.

Population Population refers to all members of a well-defined group of organisms, events, or things.

Homozygous When an organism is referred to as being homozygous for a specific gene, it means that it carries two identical copies of that gene for a given trait on the two corresponding chromosomes (e.g., the genotype is AA or aa). Such a cell or an organism is called a homozygote.

Phenotype The phenotype of an individual organism is either its total physical appearance and constitution, or a specific manifestation of a trait, such as size or eye color, that varies between individuals. Phenotype is determined to some extent by genotype, or by the identity of the alleles that an individual carries at one or more positions on the chromosomes.

Genotype The genotype is the specific genetic makeup of an individual, usually in the form of DNA. It codes for the phenotype of that individual. Any given gene will usually cause an observable change in an organism, known as the phenotype.

Trait An enduring personality characteristic that tends to lead to certain behaviors is called a trait. The term trait also means a genetically inherited feature of an organism.

Corpus callosum The corpus callosum is the largest white matter structure in the brain. It consists of mostly of contralateral axon projections. The corpus callosum connects the left and right cerebral hemispheres. Most communication between regions in different halves of the brain are carried over the corpus callosum.

Forebrain The forebrain is the highest level of the brain. Key structures in the forebrain are the limbic system, thalamus, basal ganglia, hypothalamus, and cerebral cortex.

Embryo A developed zygote that has a rudimentary heart, brain, and other organs is referred to as an embryo.

Recessive gene Recessive gene refers to an allele that causes a phenotype (visible or detectable characteristic) that is only seen in a homozygous genotype (an organism that has two copies of the same allele). Thus, both parents have to be carriers of a recessive trait in order for a child to express that trait.

Recessive allele Recessive allele refers to an unexpressed allele in a heterozygote, but is phenotypically expressed when homozygous (an organism that has two copies of the same allele).. Thus, both parents have to be carriers of a recessive trait in order for a child to express that trait.

Dominant allele The dominant allele, of two alleles for the same trait, is the one that is expressed in a heterozygote. Many traits are determined by pairs of complementary genes, each inherited from a single parent. Often when these are paired and compared, one gene will be found to effectively shut out the instructions from the other gene.

Heredity | Heredity is the transfer of characteristics from parent to offspring through their genes.

Mendel | Mendel is often called the "father of genetics" for his study of the inheritance of traits in pea plants. Mendel showed that there was particulate inheritance of traits according to his laws of inheritance.

Tay-Sachs disease | Tay-Sachs disease is a fatal genetic disorder, inherited in an autosomal recessive pattern, in which harmful quantities of a fatty substance called ganglioside GM2 accumulate in the nerve cells in the brain.

Ethnic group | An ethnic group is a culture or subculture whose members are readily distinguishable by outsiders based on traits originating from a common racial, national, linguistic, or religious source. Members of an ethnic group are often presumed to be culturally or biologically similar, although this is not in fact necessarily the case.

Chorea | Chorea is the occurrence of continuous rapid, jerky, involuntary movements that may involve the face and limb and result in an inability to maintain a posture. It is also known as St. Vitus Dance disease and is seen mostly in children.

Infancy | The developmental period that extends from birth to 18 or 24 months is called infancy.

Syndrome | The term syndrome is the association of several clinically recognizable features, signs, symptoms, phenomena or characteristics which often occur together, so that the presence of one feature indicates the presence of the others.

Trisomy | A condition wherein there are three rather than the usual pair of homologous chromosomes within the cell nucleus is referred to as trisomy.

Guthrie | The theory of learning proposed by Guthrie was based on one principle, Contiguity : A combination of stimuli which has accompanied a movement will on its recurrence tend to be followed by that movement. Prediction of behavior will always be probabilistic.

Personality | Personality refers to the pattern of enduring characteristics that differentiates a person, the patterns of behaviors that make each individual unique.

Basal ganglia | The basal ganglia are a group of nuclei in the brain associated with motor and learning functions.

Genetics | Genetics is the science of genes, heredity, and the variation of organisms.

Life span | Life span refers to the upper boundary of life, the maximum number of years an individual can live. The maximum life span of human beings is about 120 years of age. Females live an average of 6 years longer than males.

Selective breeding | Selective breeding refers to the mating of those members of a strain of animals or plants that manifest a particular characteristic, which may or may not be done deliberately, to affect the genetic makeup of future generations of that strain.

Species | Species refers to a reproductively isolated breeding population.

Immune system | The most important function of the human immune system occurs at the cellular level of the blood and tissues. The lymphatic and blood circulation systems are highways for specialized white blood cells. These cells include B cells, T cells, natural killer cells, and macrophages. All function with the primary objective of recognizing, attacking and destroying bacteria, viruses, cancer cells, and all substances seen as foreign.

Uterus | The uterus or womb is the major female reproductive organ. The main function of the uterus is to accept a fertilized ovum which becomes implanted into the endometrium, and derives nourishment from blood vessels which develop exclusively for this purpose.

Clone | A clone is a genetic, not behavioral, copy of an organism.

Embryonic stage	The embryonic stage lasts from the third through the eighth week following conception. During this stage the major organ systems undergo rapid differentiation.
Neurotransmitter	A neurotransmitter is a chemical that is used to relay, amplify and modulate electrical signals between a neurons and another cell.

Cerebrospinal fluid	A solution that fills the hollow cavities of the brain and circulates around the brain and spinal cord is called cerebrospinal fluid.
Descartes	Descartes was concerned with the sharp contrast between the certainty of mathematics and the controversial nature of philosophy, and came to believe that the sciences could be made to yield results as certain as those of mathematics. He introduced the method of rationalism for arriving at knowledge. He also saw the human condition as a competition between the body and soul, introducing the concept of dualism.
Nerve	A nerve is an enclosed, cable-like bundle of nerve fibers or axons, which includes the glia that ensheath the axons in myelin. Neurons are sometimes called nerve cells, though this term is technically imprecise since many neurons do not form nerves.
Brain	The brain controls and coordinates most movement, behavior and homeostatic body functions such as heartbeat, blood pressure, fluid balance and body temperature. Functions of the brain are responsible for cognition, emotion, memory, motor learning and other sorts of learning. The brain is primarily made up of two types of cells: glia and neurons.
Stimulus	A change in an environmental condition that elicits a response is a stimulus.
Nervous system	The body's electrochemical communication circuitry, made up of billions of neurons is a nervous system.
Neuron	The neuron is the primary cell of the nervous system. They are found in the brain, the spinal cord, in the nerves and ganglia of the peripheral nervous system. It is a specialized cell that conducts impulses through the nervous system and contains three major parts: cell body, dendrites, and an axon. It can have many dendrites but only one axon.
Luigi Galvani	Luigi Galvani experimented on the stimulation of frog's legs through electrical stimulation demonstrating the electrical nature of the nervous impulse.
Electrode	Any device used to electrically stimulate nerve tissue or to record its activity is an electrode.
Research method	The scope of the research method is to produce some new knowledge. This, in principle, can take three main forms: Exploratory research; Constructive research; and Empirical research.
Neocortex	The neocortex is part of the cerebral cortex which covers most of the surface of the cerebral hemispheres including the frontal, parietal, occipital, and temporal lobes. Often seen as the hallmark of human intelligence, the role of this structure in the brain appears to be involved in conscious thought, spatial reasoning, and sensory perception.
Dura mater	The dura mater is the tough and inflexible outermost of the three layers of the meninges surrounding the brain. The other two meninges are the pia mater and the arachnoid mater. The dura mater envelops and protects the brain and spinal cord.
Pupil	In the eye, the pupil is the opening in the middle of the iris. It appears black because most of the light entering it is absorbed by the tissues inside the eye. The size of the pupil is controlled by involuntary contraction and dilation of the iris, in order to regulate the intensity of light entering the eye. This is known as the pupillary reflex.
Penfield	Penfield treated patients with severe epilepsy by destroying nerve cells in the brain. Before operating, he stimulated the brain with electrical probes while the patients were conscious on the operating table, and observed their responses. It allowed him to create maps of sensory and motor functions.
Electroencep-alogram	Electroencephalography is the neurophysiologic measurement of the electrical activity of the brain by recording from electrodes placed on the scalp, or in the special cases on the cortex. The resulting traces are known as an electroencephalogram and represent so-called brainwaves.

Helmholtz	Helmholtz a pioneer of the new science of psychology, was a rigorous experimental physiologist and philospher. He gave us the distinction betwen sensation and peception and is well known for his theories of color perception and hearing.
Julius Bernstein	Julius Bernstein described the nerve impulse as a wave of negativity. The surface of the nerve below and above the impulse becomes negative as the impulse travels. This negativity in effect is electrical depolarization. His work made it possible to measure the duration of an impulse
Axon	An axon, or "nerve fiber," is a long slender projection of a nerve cell, or "neuron," which conducts electrical impulses away from the neuron's cell body or soma.
Microelectrode	An electrical wire so small that it can be used either to monitor the electrical activity of a single neuron or to stimulate activity within it is a microelectrode.
Ganglion	A ganglion is a tissue mass that contains the dendrites and cell bodies (or "somas") of nerve cells, in most case ones belonging to the peripheral nervous system. Within the central nervous system such a mass is often called a nucleus.
Baseline	Measure of a particular behavior or process taken before the introduction of the independent variable or treatment is called the baseline.
Ion	An ion is an atom or group of atoms with a net electric charge. The energy required to detach an electron in its lowest energy state from an atom or molecule of a gas with less net electric charge is called the ionization potential, or ionization energy.
Concentration gradient	An ion gradient is a concentration gradient of ions, it can be called an electrochemical potential gradient of ions across membranes.
Cell membrane	A component of every biological cell, the selectively permeable cell membrane is a thin and structured bilayer of phospholipid and protein molecules that envelopes the cell. It separates a cell's interior from its surroundings and controls what moves in and out.
Protein	A protein is a complex, high-molecular-weight organic compound that consists of amino acids joined by peptide bonds. It is essential to the structure and function of all living cells and viruses. Many are enzymes or subunits of enzymes.
Ion channel	An Ion channel is a pore-forming protein that help establish the small voltage gradient that exists across the membrane of all living cells, by controlling the flow of ions. They are present in the membranes that surround all biological cells.
Resting potential	The resting potential of a cell is the membrane potential that would be maintained if there were no action potentials, synaptic potentials, or other active changes in the membrane potential. In most cells the resting potential has a negative value, which by convention means that there is excess negative charge inside compared to outside.
Species	Species refers to a reproductively isolated breeding population.
Blocking	If the one of the two members of a compound stimulus fails to produce the CR due to an earlier conditioning of the other member of the compound stimulus, blocking has occurred.
Potassium channel	The potassium channel is the most common type of ion channel. They are found in most cells, and control the electrical excitability of the cell membrane. In neurons, they shape action potentials and set the resting membrane potential. They regulate cellular processes such as the secretion of hormones, so their malfunction can lead to diseases.
Hyperpolariz-tion	Hyperpolarization is any change in a cell's membrane potential that makes it more polarized. That is, hyperpolarization is an increase in the absolute value of a cell's membrane potential.
Depolarization	Depolarization is any change that decreases the absolute value of a cell's membrane

	potential. Thus, changes in membrane voltage in which the membrane potential becomes less positive or less negative are both depolarizations.
Action potential	The electrical impulse that provides the basis for the conduction of a neural impulse along an axon of a neuron is the action potential. When a biological cell or patch of membrane undergoes an action potential, or electrical excitation, the polarity of the transmembrane voltage swings rapidly from negative to positive and back.
Refractory period	Refractory period refers to a phase following firing during which a neuron is less sensitive to messages from other neurons and will not fire. In the sexual response cycle, it is a period of time following orgasm during which an individual is not responsive to sexual stimulation.
Threshold	In general, a threshold is a fixed location or value where an abrupt change is observed. In the sensory modalities, it is the minimum amount of stimulus energy necessary to elicit a sensory response.
Analogy	An analogy is a comparison between two different things, in order to highlight some form of similarity. Analogy is the cognitive process of transferring information from a particular subject to another particular subject.
Protein Channel	A protein channel is a pore-forming protein that help establish the small voltage gradient that exists across the membrane of all living cells, by controlling the flow of ions. They are present in the membranes that surround all biological cells.
Nerve impulse	A nerve impulse is a change in the electric potential of a neuron; a wave of depolarization spreads along the neuron and causes the release of a neurotransmitter.
Saltatory conduction	Saltatory conduction is a means by which action potentials are transmitted along myelinated nerve fibers. In myelinated axons, action potentials do not propagate as waves, but recur at successive nodes.
Myelin	Myelin is an electrically insulating fatty layer that surrounds the axons of many neurons, especially those in the peripheral nervous system. The main consequence of a myelin sheath is an increase in the speed at which impulses propagate along the myelinated fiber. The sheath continues to develop throughout childhood.
Glial	Glial cells are non-neuronal cells that provide support and nutrition, maintain homeostasis, form myelin, and participate in signal transmission in the nervous system.
Nodes of Ranvier	Nodes of Ranvier are regularly spaced gaps in the myelin sheath around an axon or nerve fiber. About one micrometer in length, these gaps expose the membrane of the axon to the surrounding liquid. Ion flow occurs only at the nodes of Ranvier.
Eccles	Eccles was key to a number of important developments in neuroscience. Until around 1949, Eccles believed that synaptic transmission was primarily electrical rather than chemical. Although he was wrong in this hypothesis, his arguments led himself and others to perform some of the experiments which proved chemical synaptic transmission.
Motor neuron	A motor neuron is an efferent neuron that originates in the spinal cord and synapses with muscle fibers to facilitate muscle contraction and with muscle spindles to modify proprioceptive sensitivity.
Dendritic spine	A dendritic spine is a small membranous extrusion that protrudes from a dendrite and forms one half of a synapse. Changes in dendritic spine density underlie many brain functions, including motivation, learning, and memory. In particular, long-term memory is mediated in part by the growth of new dendritic spines to reinforce a particular neural pathway.
Dendrite	A dendrite is a slender, typically branched projection of a nerve cell, or "neuron," which conducts the electrical stimulation received from other cells to the body or soma of the cell

	from which it projects. This stimulation arrives through synapses, which typically are located near the tips of the dendrites and away from the soma.
Inhibitory postsynaptic potential	Inhibitory postsynaptic potential refers to a local, graded excitatory potential that hyperpolarizes a neuron's membrane--driving the charge farther negative. Driving a cell to negativity will usually lower the firing rate.
Spinal cord	The spinal cord is a part of the vertebrate nervous system that is enclosed in and protected by the vertebral column (it passes through the spinal canal). It consists of nerve cells. The spinal cord carries sensory signals and motor innervation to most of the skeletal muscles in the body.
Sensory nerves	Sensory nerves bring impulses toward the central nervous system.
Excitatory postsynaptic potential	An excitatory postsynaptic potential is a temporary decrease in postsynaptic membrane potential caused by the flow of positively-charged ions into the postsynaptic cell.
Affect	A subjective feeling or emotional tone often accompanied by bodily expressions noticeable to others is called affect.
Gland	A gland is an organ in an animal's body that synthesizes a substance for release such as hormones, often into the bloodstream or into cavities inside the body or its outer surface.
Axon terminal	A swelling at the end of an axon that is designed to release a chemical substance onto another neuron, muscle cell, or gland cell is called the axon terminal.
Receptor	A sensory receptor is a structure that recognizes a stimulus in the internal or external environment of an organism. In response to stimuli the sensory receptor initiates sensory transduction by creating graded potentials or action potentials in the same cell or in an adjacent one.
Acetylcholine	The chemical compound acetylcholine was the first neurotransmitter to be identified. It plays a role in learning, memory, and rapid eye movement sleep and causes the skeletal muscle fibers to contract.
Synapse	A synapse is specialized junction through which cells of the nervous system signal to one another and to non-neuronal cells such as muscles or glands.
Autoimmune disease	Autoimmune disease arises from an overactive immune response of the body against substances and tissues normally present in the body. In other words, the body attacks its own cells.
Immune system	The most important function of the human immune system occurs at the cellular level of the blood and tissues. The lymphatic and blood circulation systems are highways for specialized white blood cells. These cells include B cells, T cells, natural killer cells, and macrophages. All function with the primary objective of recognizing, attacking and destroying bacteria, viruses, cancer cells, and all substances seen as foreign.
Syndrome	The term syndrome is the association of several clinically recognizable features, signs, symptoms, phenomena or characteristics which often occur together, so that the presence of one feature indicates the presence of the others.
Thyroid	In anatomy, the thyroid is the largest endocrine gland in the body. The primary function of the thyroid is production of hormones.
Axon hillock	The axon hillock is the anatomical part of a neuron that connects the cell body to the axon. It is regarded as the place where EPSPs from numerous synaptic inputs on the dendrites or cell body accumulate.
Sensory neuron	A sensory neuron is an afferent nerve cell within the nervous system responsible for converting external stimuli from the organism's environment into internal electrical

impulses. It carries messages from a sensory organ, through a nerve, into the brain or spinal cord.

Tactile Pertaining to the sense of touch is referred to as tactile.

Sensation Sensation is the first stage in the chain of biochemical and neurologic events that begins with the impinging of a stimulus upon the receptor cells of a sensory organ, which then leads to perception, the mental state that is reflected in statements like "I see a uniformly blue wall."

Sensory receptor A sensory receptor is a structure that recognizes a stimulus in the environment of an organism. In response to stimuli the sensory receptor initiates sensory transduction by creating graded potentials or action potentials in the same cell or in an adjacent one.

Displacement An unconscious defense mechanism in which the individual directs aggressive or sexual feelings away from the primary object to someone or something safe is referred to as displacement. Displacement in linguistics is simply the ability to talk about things not present.

Charcot Charcot took an interest in the malady then called hysteria. It seemed to be a mental disorder with physical manifestations, of immediate interest to a neurologist. He believed that hysteria was the result of a weak neurological system which was hereditary.

Theories Theories are logically self-consistent models or frameworks describing the behavior of a certain natural or social phenomenon. They are broad explanations and predictions concerning phenomena of interest.

Neurotransmitter A neurotransmitter is a chemical that is used to relay, amplify and modulate electrical signals between a neurons and another cell.

Receptor site A location on the dendrite of a receiving neuron that is tailored to receive a specific neurotransmitter is a receptor site.

Limbic system The limbic system is a group of brain structures that are involved in various emotions such as aggression, fear, pleasure and also in the formation of memory. The limbic system affects the endocrine system and the autonomic nervous system. It consists of several subcortical structures located around the thalamus.

Anesthesia Anesthesia is the process of blocking the perception of pain and other sensations. This allows patients to undergo surgery and other procedures without the distress and pain they would otherwise experience.

Visual cortex The visual cortex is the general term applied to both the primary visual cortex and the visual association area. Anatomically, the visual cortex occupies the entire occipital lobe, the inferior temporal lobe (IT), posterior parts of the parietal lobe, and a few small regions in the frontal lobe.

Deep sleep Deep sleep refers to stage 4 sleep; the deepest form of normal sleep.

Amplitude Amplitude is a nonnegative scalar measure of a wave's magnitude of oscillation.

Epilepsy Epilepsy is a chronic neurological condition characterized by recurrent unprovoked neural discharges. It is commonly controlled with medication, although surgical methods are used as well.

Anatomy Anatomy is the branch of biology that deals with the structure and organization of living things. It can be divided into animal anatomy (zootomy) and plant anatomy (phytonomy). Major branches of anatomy include comparative anatomy, histology, and human anatomy.

Seizure A seizure is a temporary alteration in brain function expressed as a changed mental state, tonic or clonic movements and various other symptoms. They are due to temporary abnormal

	electrical activity of a group of brain cells.
Stages	Stages represent relatively discrete periods of time in which functioning is qualitatively different from functioning at other periods.
Neuropsychology	Neuropsychology is a branch of psychology that aims to understand how the structure and function of the brain relates to specific psychological processes.
Alpha wave	The brain wave associated with deep relaxation is referred to as the alpha wave. Recorded by electroencephalography (EEG) , they are synchronous and coherent (regular like sawtooth) and in the frequency range of 8 - 12 Hz. It is also called Berger's wave in memory of the founder of EEG.
Transcendental Meditation	The simplified form of meditation brought to the United States by the Maharishi Mahesh Yogi and used as a method for coping with stress is called transcendental meditation.
Learning	Learning is a relatively permanent change in behavior that results from experience. Thus, to attribute a behavioral change to learning, the change must be relatively permanent and must result from experience.
Consciousness	The awareness of the sensations, thoughts, and feelings being experienced at a given moment is called consciousness.
Trauma	Trauma refers to a severe physical injury or wound to the body caused by an external force, or a psychological shock having a lasting effect on mental life.
Tumor	A tumor is an abnormal growth that when located in the brain can either be malignant and directly destroy brain tissue, or be benign and disrupt functioning by increasing intracranial pressure.
Decoding	Process of phonetic analysis by which a printed word is converted to spoken form before retrieval from long-term memory is called decoding.
Postsynaptic Potential	A postsynaptic potential is a function of two neurons synapsing, whereby Neuron A causes a change in Neuron B's membrane potential.
Psychiatrist	A psychiatrist is a physician who specializes in the diagnosis and treatment of psychological disorders.

Excitatory message	An excitatory message is a chemical message that makes it more likely that a receiving neuron will fire and an action potential will travel down its axons.
Inhibitory message	A chemical message that prevents a receiving neuron from firing is called an inhibitory message.
Neuron	The neuron is the primary cell of the nervous system. They are found in the brain, the spinal cord, in the nerves and ganglia of the peripheral nervous system. It is a specialized cell that conducts impulses through the nervous system and contains three major parts: cell body, dendrites, and an axon. It can have many dendrites but only one axon.
Mand	The mand is verbal behavior whose form is controlled by states of deprivation and aversion; it is often said to "specify its own reinforcer." What this means loosely is that the function of a mand is to request or to obtain what is wanted.
Vagus nerve	The vagus nerve is tenth of twelve paired cranial nerves and is the only nerve that starts in the brainstem (somewhere in the medulla oblongata) and extends all the way down past the head, right down to the abdomen. The vagus nerve is arguably the single most important nerve in the body.
Acetylcholine	The chemical compound acetylcholine was the first neurotransmitter to be identified. It plays a role in learning, memory, and rapid eye movement sleep and causes the skeletal muscle fibers to contract.
Nerve	A nerve is an enclosed, cable-like bundle of nerve fibers or axons, which includes the glia that ensheath the axons in myelin. Neurons are sometimes called nerve cells, though this term is technically imprecise since many neurons do not form nerves.
Epinephrine	Epinephrine is a hormone and a neurotransmitter. Epinephrine plays a central role in the short-term stress reaction—the physiological response to threatening or exciting conditions. It is secreted by the adrenal medulla. When released into the bloodstream, epinephrine binds to multiple receptors and has numerous effects throughout the body.
Neurotransmitter	A neurotransmitter is a chemical that is used to relay, amplify and modulate electrical signals between a neurons and another cell.
Case study	A carefully drawn biography that may be obtained through interviews, questionnaires, and psychological tests is called a case study.
Stages	Stages represent relatively discrete periods of time in which functioning is qualitatively different from functioning at other periods.
Tremor	Tremor is the rhythmic, oscillating shaking movement of the whole body or just a certain part of it, caused by problems of the neurons responsible from muscle action.
Charcot	Charcot took an interest in the malady then called hysteria. It seemed to be a mental disorder with physical manifestations, of immediate interest to a neurologist. He believed that hysteria was the result of a weak neurological system which was hereditary.
Substantia nigra	The substantia nigra is a portion of the midbrain thought to be involved in certain aspects of movement and attention. Degeneration of cells in this region is the principle pathology that underlies Parkinson's disease.
Brain	The brain controls and coordinates most movement, behavior and homeostatic body functions such as heartbeat, blood pressure, fluid balance and body temperature. Functions of the brain are responsible for cognition, emotion, memory, motor learning and other sorts of learning. The brain is primarily made up of two types of cells: glia and neurons.
Basal ganglia	The basal ganglia are a group of nuclei in the brain associated with motor and learning functions.

Dopamine	Dopamine is critical to the way the brain controls our movements and is a crucial part of the basal ganglia motor loop. It is commonly associated with the 'pleasure system' of the brain, providing feelings of enjoyment and reinforcement to motivate us to do, or continue doing, certain activities.
Axon	An axon, or "nerve fiber," is a long slender projection of a nerve cell, or "neuron," which conducts electrical impulses away from the neuron's cell body or soma.
Synapse	A synapse is specialized junction through which cells of the nervous system signal to one another and to non-neuronal cells such as muscles or glands.
Dendrite	A dendrite is a slender, typically branched projection of a nerve cell, or "neuron," which conducts the electrical stimulation received from other cells to the body or soma of the cell from which it projects. This stimulation arrives through synapses, which typically are located near the tips of the dendrites and away from the soma.
Receptor	A sensory receptor is a structure that recognizes a stimulus in the internal or external environment of an organism. In response to stimuli the sensory receptor initiates sensory transduction by creating graded potentials or action potentials in the same cell or in an adjacent one.
Dendritic spine	A dendritic spine is a small membranous extrusion that protrudes from a dendrite and forms one half of a synapse. Changes in dendritic spine density underlie many brain functions, including motivation, learning, and memory. In particular, long-term memory is mediated in part by the growth of new dendritic spines to reinforce a particular neural pathway.
Axon terminal	A swelling at the end of an axon that is designed to release a chemical substance onto another neuron, muscle cell, or gland cell is called the axon terminal.
Protein	A protein is a complex, high-molecular-weight organic compound that consists of amino acids joined by peptide bonds. It is essential to the structure and function of all living cells and viruses. Many are enzymes or subunits of enzymes.
Receptor site	A location on the dendrite of a receiving neuron that is tailored to receive a specific neurotransmitter is a receptor site.
Mitochondria	Mitochondria are sometimes described as "cellular power plants", because their primary function is to convert organic materials into energy in the form of ATP.
Organelle	An organelle is one of several structures with specialized functions, suspended in the cytoplasm of a eukaryotic cell.
Glial	Glial cells are non-neuronal cells that provide support and nutrition, maintain homeostasis, form myelin, and participate in signal transmission in the nervous system.
Nervous system	The body's electrochemical communication circuitry, made up of billions of neurons is a nervous system.
Synaptic cleft	Synaptic cleft refers to a microscopic gap between the terminal button of a neuron and the cell membrane of another neuron.
Depolarization	Depolarization is any change that decreases the absolute value of a cell's membrane potential. Thus, changes in membrane voltage in which the membrane potential becomes less positive or less negative are both depolarizations.
Postsynaptic Potential	A postsynaptic potential is a function of two neurons synapsing, whereby Neuron A causes a change in Neuron B's membrane potential.
Action potential	The electrical impulse that provides the basis for the conduction of a neural impulse along an axon of a neuron is the action potential. When a biological cell or patch of membrane undergoes an action potential, or electrical excitation, the polarity of the transmembrane

	voltage swings rapidly from negative to positive and back.
Reuptake	Reuptake is the reabsorption of a neurotransmitter by the molecular transporter of a pre-synaptic neuron after it has performed its function of transmitting a neural impulse.
Enzyme	An enzyme is a protein that catalyzes, or speeds up, a chemical reaction. Enzymes are essential to sustain life because most chemical reactions in biological cells would occur too slowly, or would lead to different products, without enzymes.
Capillary	The capillary is the smallest of a body's blood vessels, measuring 5-10 ìm. They connect arteries and veins, and most closely interact with tissues.
Central nervous system	The vertebrate central nervous system consists of the brain and spinal cord.
Axon hillock	The axon hillock is the anatomical part of a neuron that connects the cell body to the axon. It is regarded as the place where EPSPs from numerous synaptic inputs on the dendrites or cell body accumulate.
Blocking	If the one of the two members of a compound stimulus fails to produce the CR due to an earlier conditioning of the other member of the compound stimulus, blocking has occurred.
Neurologist	A physician who studies the nervous system, especially its structure, functions, and abnormalities is referred to as neurologist.
Peripheral nervous system	The peripheral nervous system consists of the nerves and neurons that serve the limbs and organs. It is not protected by bone or the blood-brain barrier, leaving it exposed to toxins and mechanical injuries. The peripheral nervous system is divided into the somatic nervous system and the autonomic nervous system.
Spinal cord	The spinal cord is a part of the vertebrate nervous system that is enclosed in and protected by the vertebral column (it passes through the spinal canal). It consists of nerve cells. The spinal cord carries sensory signals and motor innervation to most of the skeletal muscles in the body.
Motor neuron	A motor neuron is an efferent neuron that originates in the spinal cord and synapses with muscle fibers to facilitate muscle contraction and with muscle spindles to modify proprioceptive sensitivity.
Microelectrode	An electrical wire so small that it can be used either to monitor the electrical activity of a single neuron or to stimulate activity within it is a microelectrode.
Electrode	Any device used to electrically stimulate nerve tissue or to record its activity is an electrode.
Staining	Staining is a biochemical technique of adding a class-specific (DNA, proteins, lipids, carbohydrates) dye to a substrate to qualify or quantify the presence of a specific compound. They are frequently used to highlight structures in tissues for viewing, often with the aid of different microscopes.
Interneuron	An interneuron (also called relay neuron or association neuron) is a neuron that communicates only to other neurons.
Feedback	Feedback refers to information returned to a person about the effects a response has had.
Neuropeptides	Brain chemicals, such as enkephalins and endorphins, that regulate the activity of neurons are called neuropeptides.
Amino acid	Amino acid is the basic structural building unit of proteins. They form short polymer chains called peptides or polypeptides which in turn form structures called proteins.
Tyrosine	Tyrosine is one of the 20 amino acids that are used by cells to synthesize proteins. It plays

	a key role in signal transduction, since it can be tagged (phosphorylated) with a phosphate group by protein kinases to alter the functionality and activity of certain enzymes.
Norepinephrine	Norepinephrine is released from the adrenal glands as a hormone into the blood, but it is also a neurotransmitter in the nervous system. As a stress hormone, it affects parts of the human brain where attention and impulsivity are controlled. Along with epinephrine, this compound effects the fight-or-flight response, activating the sympathetic nervous system to directly increase heart rate, release energy from fat, and increase muscle readiness.
Glutamate	Glutamate is one of the 20 standard amino acids used by all organisms in their proteins. It is critical for proper cell function, but it is not an essential nutrient in humans because it can be manufactured from other compounds.
Forebrain	The forebrain is the highest level of the brain. Key structures in the forebrain are the limbic system, thalamus, basal ganglia, hypothalamus, and cerebral cortex.
Ribosome	A ribosome is an organelle composed of rRNA and ribosomal proteins. It translates mRNA into a polypeptide chain (e.g., a protein). It can be thought of as a factory that builds a protein from a set of genetic instructions.
Golgi	Golgi discovered a method of staining nervous tissue which would stain a limited number of cells at random, in their entirety. This enabled him to view the paths of nerve cells in the brain for the first time. He called his discovery the black reaction. It is now known universally as the Golgi stain.
Peptides	Peptides are the family of molecules formed from the linking, in a defined order, of various amino acids.
Dynorphin	Dynorphin is a popular and powerful opioid ligand. It may act as an antidote to pleasurable effects of cocaine. As such, it may help some individuals against the trap of addiction. Blocking it may help alleviate depression. Dynorphin is classified as a kappa opioid receptor agonist.
Oxytocin	Oxytocin is synthesized in magnocellular neurosecretory cells in the hypothalamus and released by the posterior lobe of the pituitary gland. It is involved in the facilitation of birth and breastfeeding as well as in bonding.
Learning	Learning is a relatively permanent change in behavior that results from experience. Thus, to attribute a behavioral change to learning, the change must be relatively permanent and must result from experience.
Hormone	A hormone is a chemical messenger from one cell (or group of cells) to another. The best known are those produced by endocrine glands, but they are produced by nearly every organ system. The function of hormones is to serve as a signal to the target cells; the action of the hormone is determined by the pattern of secretion and the signal transduction of the receiving tissue.
Opioid	An opioid is any agent that binds to opioid receptors, found principally in the central nervous system and gastrointestinal tract.
Opium	Opium is a narcotic analgesic drug which is obtained from the unripe seed pods of the opium poppy. Regular use, even for a few days, invariably leads to physical tolerance and dependence. Various degrees of psychological addiction can occur, though this is relatively rare when opioids are properly used..
Morphine	Morphine, the principal active agent in opium, is a powerful opioid analgesic drug. According to recent research, it may also be produced naturally by the human brain. Morphine is usually highly addictive, and tolerance and physical and psychological dependence develop quickly.
Ion channel	An Ion channel is a pore-forming protein that help establish the small voltage gradient that

	exists across the membrane of all living cells, by controlling the flow of ions. They are present in the membranes that surround all biological cells.
Cell membrane	A component of every biological cell, the selectively permeable cell membrane is a thin and structured bilayer of phospholipid and protein molecules that envelopes the cell. It separates a cell's interior from its surroundings and controls what moves in and out.
Ion	An ion is an atom or group of atoms with a net electric charge. The energy required to detach an electron in its lowest energy state from an atom or molecule of a gas with less net electric charge is called the ionization potential, or ionization energy.
Brightness	The dimension of visual sensation that is dependent on the intensity of light reflected from a surface and that corresponds to the amplitude of the light wave is called brightness.
Affect	A subjective feeling or emotional tone often accompanied by bodily expressions noticeable to others is called affect.
Cytoplasm	Cytoplasm is the colloidal, semi-fluid matter contained within the cell's plasma membrane, in which organelles are suspended. In contrast to the protoplasm, the cytoplasm does not include the cell nucleus, the interior of which is made up of nucleoplasm.
Skeletal muscle	Skeletal muscle is a type of striated muscle, attached to the skeleton. They are used to facilitate movement, by applying force to bones and joints; via contraction. They generally contract voluntarily (via nerve stimulation), although they can contract involuntarily.
Nicotine	Nicotine is an organic compound, an alkaloid found naturally throughout the tobacco plant, with a high concentration in the leaves. It is a potent nerve poison and is included in many insecticides. In lower concentrations, the substance is a stimulant and is one of the main factors leading to the pleasure and habit-forming qualities of tobacco smoking.
Autonomic nervous system	A division of the peripheral nervous system, the autonomic nervous system, regulates glands and activities such as heartbeat, respiration, digestion, and dilation of the pupils. It is responsible for homeostasis, maintaining a relatively constant internal environment .
Parasympathetic	The parasympathetic nervous system is one of two divisions of the autonomic nervous system. It conserves energy as it slows the heart rate, increases intestinal and gland activity, and relaxes sphincter muscles. In another words, it acts to reverse the effects of the Sympathetic nervous system.
Sympathetic	The sympathetic nervous system activates what is often termed the "fight or flight response". It is an automatic regulation system, that is, one that operates without the intervention of conscious thought.
Adrenaline	Adrenaline refers to a hormone produced by the adrenal medulla that stimulates sympathetic ANS activity and generally arouses people and heightens their emotional responsiveness.
Nucleus	In neuroanatomy, a cluster of cell bodies of neurons within the central nervous system is a nucleus.
Serotonin	Serotonin, a neurotransmitter, is believed to play an important part of the biochemistry of depression, bipolar disorder and anxiety. It is also believed to be influential on sexuality and appetite.
Neocortex	The neocortex is part of the cerebral cortex which covers most of the surface of the cerebral hemispheres including the frontal, parietal, occipital, and temporal lobes. Often seen as the hallmark of human intelligence, the role of this structure in the brain appears to be involved in conscious thought, spatial reasoning, and sensory perception.
Midbrain	Located between the hindbrain and forebrain, a region in which many nerve-fiber systems ascend and descend to connect the higher and lower portions of the brain is referred to as

	midbrain. It is archipallian in origin, meaning its general architecture is shared with the most ancient of vertebrates. Dopamine produced in the subtantia nigra plays a role in motivation and habituation of species from humans to the most elementary animals such as insects.
Locus coeruleus	The Locus coeruleus is a nucleus in the brain stem (inferior to the cerebellum in the caudal midbrain/rostral pons) apparently responsible for the physiological reactions involved in stress and panic.
Depression	In everyday language depression refers to any downturn in mood, which may be relatively transitory and perhaps due to something trivial. This is differentiated from Clinical depression which is marked by symptoms that last two weeks or more and are so severe that they interfere with daily living.
Mania	Mania is a medical condition characterized by severely elevated mood. Mania is most usually associated with bipolar disorder, where episodes of mania may cyclically alternate with episodes of depression.
Schizophrenia	Schizophrenia is characterized by persistent defects in the perception or expression of reality. A person suffering from untreated schizophrenia typically demonstrates grossly disorganized thinking, and may also experience delusions or auditory hallucinations
Tics	Tics are a repeated, impulsive action, almost reflexive in nature, which the person feels powerless to control or avoid.
Cerebellum	The cerebellum is located in the inferior posterior portion of the head (the hindbrain), directly dorsal to the brainstem and pons, inferior to the occipital lobe. The cerebellum is a region of the brain that plays an important role in the integration of sensory perception and fine motor output.
Hippocampus	The hippocampus is a part of the brain located inside the temporal lobe. It forms a part of the limbic system and plays a part in memory and navigation.
Heroin	Heroin is widely and illegally used as a powerful and addictive drug producing intense euphoria, which often disappears with increasing tolerance. Heroin is a semi-synthetic opioid. It is the 3,6-diacetyl derivative of morphine and is synthesised from it by acetylation.
Plasticity	The capacity for modification and change is referred to as plasticity.
Hebb	Hebb demonstrated that the rearing of rats in an enriched environment could alter neural development and that sensory - neural connections were shaped by experience. He is famous for developing the concept of neural nets. He also believed that learning early in life is of the incremental variety, whereas later it is cognitive, insightful, and more all-or-none.
Habituation	In habituation there is a progressive reduction in the response probability with continued repetition of a stimulus.
Stimulus	A change in an environmental condition that elicits a response is a stimulus.
Senses	The senses are systems that consist of a sensory cell type that respond to a specific kind of physical energy, and that correspond to a defined region within the brain where the signals are received and interpreted.
Sensation	Sensation is the first stage in the chain of biochemical and neurologic events that begins with the impinging of a stimulus upon the receptor cells of a sensory organ, which then leads to perception, the mental state that is reflected in statements like "I see a uniformly blue wall."
Sensory neuron	A sensory neuron is an afferent nerve cell within the nervous system responsible for

	converting external stimuli from the organism's environment into internal electrical impulses. It carries messages from a sensory organ, through a nerve, into the brain or spinal cord.
Excitatory postsynaptic potential	An excitatory postsynaptic potential is a temporary decrease in postsynaptic membrane potential caused by the flow of positively-charged ions into the postsynaptic cell.
Sensitization	Sensitization is a process whereby an organism is made more responsive to certain aspects of its environment. For example, increases in the effects of a drug as a result of repeated administration. Also known as reverse tolerance.
Potassium channel	The potassium channel is the most common type of ion channel. They are found in most cells, and control the electrical excitability of the cell membrane. In neurons, they shape action potentials and set the resting membrane potential. They regulate cellular processes such as the secretion of hormones, so their malfunction can lead to diseases.
Amplitude	Amplitude is a nonnegative scalar measure of a wave's magnitude of oscillation.
Associative learning	A metal connection made between two events is called associative learning.
Nerve impulse	A nerve impulse is a change in the electric potential of a neuron; a wave of depolarization spreads along the neuron and causes the release of a neurotransmitter.
Lesion	A lesion is a non-specific term referring to abnormal tissue in the body. It can be caused by any disease process including trauma (physical, chemical, electrical), infection, neoplasm, metabolic and autoimmune.
Mutation	Mutation is a permanent, sometimes transmissible (if the change is to a germ cell) change to the genetic material (usually DNA or RNA) of a cell. They can be caused by copying errors in the genetic material during cell division and by exposure to radiation, chemicals, or viruses, or can occur deliberately under cellular control during the processes such as meiosis or hypermutation.
Ganglion	A ganglion is a tissue mass that contains the dendrites and cell bodies (or "somas") of nerve cells, in most case ones belonging to the peripheral nervous system. Within the central nervous system such a mass is often called a nucleus.
Research method	The scope of the research method is to produce some new knowledge. This, in principle, can take three main forms: Exploratory research; Constructive research; and Empirical research.
Insight	Insight refers to a sudden awareness of the relationships among various elements that had previously appeared to be independent of one another.
Reinforcement	In operant conditioning, reinforcement is any change in an environment that (a) occurs after the behavior, (b) seems to make that behavior re-occur more often in the future and (c) that reoccurence of behavior must be the result of the change.
Excitatory synapse	An excitatory synapse is where the neurotransmitter increases the likelihood that an action potential will occur, or increases the rate at which they are already occurring, in the neuron on which it acts.
Inhibitory synapse	An inhibitory synapse is a synapse in which an action potential in the presynaptic cell decreases the probability of an action potential occurring in the postsynaptic cell.
Gamma-aminobutyric acid	Gamma-aminobutyric acid is a neurotransmitter that reduces activity across the synapse and thus inhibits a range of behaviors and emotions, especially generalized anxiety.

Hippocampus	The hippocampus is a part of the brain located inside the temporal lobe. It forms a part of the limbic system and plays a part in memory and navigation.
Amygdala	Located in the brain's medial temporal lobe, the almond-shaped amygdala is believed to play a key role in the emotions. It forms part of the limbic system and is linked to both fear responses and pleasure. Its size is positively correlated with aggressive behavior across species.
Thalamus	An area near the center of the brain involved in the relay of sensory information to the cortex and in the functions of sleep and attention is the thalamus.
Hormone	A hormone is a chemical messenger from one cell (or group of cells) to another. The best known are those produced by endocrine glands, but they are produced by nearly every organ system. The function of hormones is to serve as a signal to the target cells; the action of the hormone is determined by the pattern of secretion and the signal transduction of the receiving tissue.
Brain	The brain controls and coordinates most movement, behavior and homeostatic body functions such as heartbeat, blood pressure, fluid balance and body temperature. Functions of the brain are responsible for cognition, emotion, memory, motor learning and other sorts of learning. The brain is primarily made up of two types of cells: glia and neurons.
Neurotransmitter	A neurotransmitter is a chemical that is used to relay, amplify and modulate electrical signals between a neurons and another cell.
Receptor	A sensory receptor is a structure that recognizes a stimulus in the internal or external environment of an organism. In response to stimuli the sensory receptor initiates sensory transduction by creating graded potentials or action potentials in the same cell or in an adjacent one.
Psychoactive drug	A psychoactive drug or psychotropic substance is a chemical that alters brain function, resulting in temporary changes in perception, mood, consciousness, or behavior. Such drugs are often used for recreational and spiritual purposes, as well as in medicine, especially for treating neurological and psychological illnesses.
Nervous system	The body's electrochemical communication circuitry, made up of billions of neurons is a nervous system.
Lungs	The lungs are the essential organs of respiration. Its principal function is to transport oxygen from the atmosphere into the bloodstream, and excrete carbon dioxide from the bloodstream into the atmosphere.
Endothelial cell	An endothelial cell forms an interface between circulating blood in the lumen and the rest of the vessel wall. They line the entire circulatory system, from the heart to the smallest capillary. In small blood vessels and capillaries, they are often the only cell-type present.
Capillary	The capillary is the smallest of a body's blood vessels, measuring 5-10 ìm. They connect arteries and veins, and most closely interact with tissues.
Astrocyte	An astrocyte, also known as astroglia, is a characteristic star-shaped cell in the brain. They are the biggest cells found in brain tissue and outnumber the neurons ten to one. A commonly accepted function is to structure physically the brain. A second function is to provide neurons with nutrients such as glucose. They regulate the flow of nutrients provided by capillaries by forming the blood-brain barrier.
Glial	Glial cells are non-neuronal cells that provide support and nutrition, maintain homeostasis, form myelin, and participate in signal transmission in the nervous system.
Amino acid	Amino acid is the basic structural building unit of proteins. They form short polymer chains called peptides or polypeptides which in turn form structures called proteins.

Glucose	Glucose, a simple monosaccharide sugar, is one of the most important carbohydrates and is used as a source of energy in animals and plants. Glucose is one of the main products of photosynthesis and starts respiration.
Blood-brain barrier	The blood-brain barrier is a physical barrier between the blood vessels in the central nervous system, and most parts of the central nervous system itself. The blood-brain barrier appears to exist primarily to protect the brain from the chemical messenger systems flowing around the body.
Pituitary gland	The pituitary gland is an endocrine gland about the size of a pea that sits in the small, bony cavity at the base of the brain. The pituitary gland secretes hormones regulating a wide variety of bodily activities, including trophic hormones that stimulate other endocrine glands.
Hypothalamus	The hypothalamus is a region of the brain located below the thalamus, forming the major portion of the ventral region of the diencephalon and functioning to regulate certain metabolic processes and other autonomic activities.
Pineal gland	The pineal gland is a small endocrine gland. It is located near the center of the brain, between the two hemispheres and near the central switching point of the thalamic bodies. It is responsible for the production of melatonin, which has a role in regulating circadian rhythms.
Neuron	The neuron is the primary cell of the nervous system. They are found in the brain, the spinal cord, in the nerves and ganglia of the peripheral nervous system. It is a specialized cell that conducts impulses through the nervous system and contains three major parts: cell body, dendrites, and an axon. It can have many dendrites but only one axon.
Pathogen	A pathogen or infectious agent is a biological agent that causes disease or illness to its host.
Central nervous system	The vertebrate central nervous system consists of the brain and spinal cord.
Dopamine	Dopamine is critical to the way the brain controls our movements and is a crucial part of the basal ganglia motor loop. It is commonly associated with the 'pleasure system' of the brain, providing feelings of enjoyment and reinforcement to motivate us to do, or continue doing, certain activities.
Affect	A subjective feeling or emotional tone often accompanied by bodily expressions noticeable to others is called affect.
Liver	The liver plays a major role in metabolism and has a number of functions in the body including detoxification, glycogen storage and plasma protein synthesis. It also produces bile, which is important for digestion. The liver converts most carbohydrates, proteing, and fats into glucose.
Individual differences	Individual differences psychology studies the ways in which individual people differ in their behavior. This is distinguished from other aspects of psychology in that although psychology is ostensibly a study of individuals, modern psychologists invariably study groups.
Synapse	A synapse is specialized junction through which cells of the nervous system signal to one another and to non-neuronal cells such as muscles or glands.
Axon	An axon, or "nerve fiber," is a long slender projection of a nerve cell, or "neuron," which conducts electrical impulses away from the neuron's cell body or soma.
Agonist	Agonist refers to a drug that mimics or increases a neurotransmitter's effects.
Reuptake	Reuptake is the reabsorption of a neurotransmitter by the molecular transporter of a pre-

	synaptic neuron after it has performed its function of transmitting a neural impulse.
Acetylcholine	The chemical compound acetylcholine was the first neurotransmitter to be identified. It plays a role in learning, memory, and rapid eye movement sleep and causes the skeletal muscle fibers to contract.
Nicotine	Nicotine is an organic compound, an alkaloid found naturally throughout the tobacco plant, with a high concentration in the leaves. It is a potent nerve poison and is included in many insecticides. In lower concentrations, the substance is a stimulant and is one of the main factors leading to the pleasure and habit-forming qualities of tobacco smoking.
Curare	Curare is a drug that once entered into the bloodstream and reaches the muscles it blocks the receptors on the muscles, paralyzing the muscles.
Skeletal muscle	Skeletal muscle is a type of striated muscle, attached to the skeleton. They are used to facilitate movement, by applying force to bones and joints; via contraction. They generally contract voluntarily (via nerve stimulation), although they can contract involuntarily.
Enzyme	An enzyme is a protein that catalyzes, or speeds up, a chemical reaction. Enzymes are essential to sustain life because most chemical reactions in biological cells would occur too slowly, or would lead to different products, without enzymes.
Stimulant	A stimulant is a drug which increases the activity of the sympathetic nervous system and produces a sense of euphoria or awakeness.
Narcotic	The term narcotic originally referred to a variety of substances that induced sleep (such state is narcosis). In legal context, narcotic refers to opium, opium derivatives, and their semisynthetic or totally synthetic substitutes.
Codeine	Codeine is an opioid used for its analgesic, antitussive and antidiarrheal properties
Heroin	Heroin is widely and illegally used as a powerful and addictive drug producing intense euphoria, which often disappears with increasing tolerance. Heroin is a semi-synthetic opioid. It is the 3,6-diacetyl derivative of morphine and is synthesised from it by acetylation.
Hallucinogen	Certain drugs can affect the subjective qualities of perception, thought or emotion, resulting in altered interpretations of sensory input, alternate states of consciousness, or hallucinations. The term hallucinogen is often broadly applied, especially in current scientific literature, to some or all of these substances.
Amphetamine	Amphetamine is a synthetic stimulant used to suppress the appetite, control weight, and treat disorders including narcolepsy and ADHD. It is also used recreationally and for performance enhancement.
Psychedelic	A psychedelic experience is characterized by the perception of aspects of one's mind previously unknown, or by the creative exuberance of the mind.
Cocaine	Cocaine is a crystalline tropane alkaloid that is obtained from the leaves of the coca plant. It is a stimulant of the central nervous system and an appetite suppressant, creating what has been described as a euphoric sense of happiness and increased energy.
Scheme	According to Piaget, a hypothetical mental structure that permits the classification and organization of new information is called a scheme.
Antidepressants	Antidepressants are medications used primarily in the treatment of clinical depression. Antidepressants create little if any immediate change in mood and require between several days and several weeks to take effect.
Mental illness	Mental illness is the term formerly used to mean psychological disorder but less preferred because it implies that the causes of the disorder can be found in a medical disease process.

Antidepressant	An antidepressant is a medication used primarily in the treatment of clinical depression. They are not thought to produce tolerance, although sudden withdrawal may produce adverse effects. They create little if any immediate change in mood and require between several days and several weeks to take effect.
Thorazine	Thorazine is the trade name for chlorpromazine, one of the antipsychotic drugs and a member of the phenothiazine group. It is classified as a low-potency antipsychotic and is used in the treatment of both acute and chronic psychoses, including schizophrenia and the manic phase of manic depression.
Sedative	A sedative is a drug that depresses the central nervous system (CNS), which causes calmness, relaxation, reduction of anxiety, sleepiness, slowed breathing, slurred speech, staggering gait, poor judgment, and slow, uncertain reflexes.
Lithium	Lithium salts are used as mood stabilizing drugs primarily in the treatment of bipolar disorder, depression, and mania; but also in treating schizophrenia. Lithium is widely distributed in the central nervous system and interacts with a number of neurotransmitters and receptors, decreasing noradrenaline release and increasing serotonin synthesis.
Kuhn	Kuhn is most famous for his book The Structure of Scientific Revolutions in which he presented the idea that science does not evolve gradually toward truth, but instead undergoes periodic revolutions which he calls "paradigm shifts."
Anesthesia	Anesthesia is the process of blocking the perception of pain and other sensations. This allows patients to undergo surgery and other procedures without the distress and pain they would otherwise experience.
Anxiety	Anxiety is a complex combination of the feeling of fear, apprehension and worry often accompanied by physical sensations such as palpitations, chest pain and/or shortness of breath.
Benzodiazepines	The benzodiazepines are a class of drugs with hypnotic, anxiolytic, anticonvulsant, amnestic and muscle relaxant properties. Benzodiazepines are often used for short-term relief of severe, disabling anxiety or insomnia.
Barbiturate	A barbiturate is a drug that acts as a central nervous system (CNS) depressant, and by virtue of this produces a wide spectrum of effects, from mild sedation to anesthesia.
Tranquilizer	A sedative, or tranquilizer, is a drug that depresses the central nervous system (CNS), which causes calmness, relaxation, reduction of anxiety, sleepiness, slowed breathing, slurred speech, staggering gait, poor judgment, and slow, uncertain reflexes.
Cross-tolerance	Cross-tolerance refers to condition in which a person may replace addiction to one drug with addiction to the other when the two drugs have similar chemical makeup and act on the same neurotransmitter receptors.
Alcoholic	An alcoholic is dependent on alcohol as characterized by craving, loss of control, physical dependence and withdrawal symptoms, and tolerance.
Syndrome	The term syndrome is the association of several clinically recognizable features, signs, symptoms, phenomena or characteristics which often occur together, so that the presence of one feature indicates the presence of the others.
Binge drinking	Binge drinking refers to consuming 5 or more drinks in a short time or drinking alchohol for the sole purpose of intoxication.
Fetus	A fetus develops from the end of the 8th week of pregnancy (when the major structures have formed), until birth.
Fetal alcohol	A cluster of abnormalities that appears in the offspring of mothers who drink alcohol heavily

syndrome	during pregnancy is called fetal alcohol syndrome.
Prenatal	Prenatal period refers to the time from conception to birth.
Receptor site	A location on the dendrite of a receiving neuron that is tailored to receive a specific neurotransmitter is a receptor site.
Ion	An ion is an atom or group of atoms with a net electric charge. The energy required to detach an electron in its lowest energy state from an atom or molecule of a gas with less net electric charge is called the ionization potential, or ionization energy.
Antianxiety drugs	Drugs that can reduce a person's level of excitability while increasing feelings of well-being are called antianxiety drugs.
Steroid	A steroid is a lipid characterized by a carbon skeleton with four fused rings. Different steroids vary in the functional groups attached to these rings. Hundreds of distinct steroids have been identified in plants and animals. Their most important role in most living systems is as hormones.
Psychiatrist	A psychiatrist is a physician who specializes in the diagnosis and treatment of psychological disorders.
Depression	In everyday language depression refers to any downturn in mood, which may be relatively transitory and perhaps due to something trivial. This is differentiated from Clinical depression which is marked by symptoms that last two weeks or more and are so severe that they interfere with daily living.
Monoamine oxidase	Monoamine oxidase is an enzyme that catalyzes the oxidation of monoamines. They are found bound to the outer membrane of mitochondria in most cell types in the body. Because of the vital role that it play in the inactivation of neurotransmitters, dysfunction (too much/too little MAO activity) is thought to be responsible for a number of neurological disorders.
Stroke	A stroke occurs when the blood supply to a part of the brain is suddenly interrupted by occlusion, by hemorrhage, or other causes
Suicide	Suicide behavior is rare in childhood but escalates in adolescence. The suicide rate increases in a linear fashion from adolescence through late adulthood.
Obesity	The state of being more than 20 percent above the average weight for a person of one's height is called obesity.
Tricyclic antidepressant	A tricyclic antidepressant is of a class of antidepressant drugs first used in the 1950s. They are named after the drugs' molecular structure, which contains three rings of atoms.
Abnormal psychology	The scientific study whose objectives are to describe, explain, predict, and control behaviors that are considered strange or unusual is referred to as abnormal psychology.
Noradrenaline	Noradrenaline is released from the adrenal glands as a hormone into the blood, but it is also a neurotransmitter in the nervous system. As a stress hormone, it affects parts of the human brain where attention and impulsivity are controlled. Along with epinephrine, this compound effects the fight-or-flight response, activating the sympathetic nervous system to directly increase heart rate, release energy from fat, and increase muscle readiness.
Serotonin	Serotonin, a neurotransmitter, is believed to play an important part of the biochemistry of depression, bipolar disorder and anxiety. It is also believed to be influential on sexuality and appetite.
Tricyclic	Tricyclic antidepressants are a class of antidepressant drugs first used in the 1950s. They are named after the drugs' molecular structure, which contains three rings of atoms.
Axon terminal	A swelling at the end of an axon that is designed to release a chemical substance onto

	another neuron, muscle cell, or gland cell is called the axon terminal.
Action potential	The electrical impulse that provides the basis for the conduction of a neural impulse along an axon of a neuron is the action potential. When a biological cell or patch of membrane undergoes an action potential, or electrical excitation, the polarity of the transmembrane voltage swings rapidly from negative to positive and back.
Blocking	If the one of the two members of a compound stimulus fails to produce the CR due to an earlier conditioning of the other member of the compound stimulus, blocking has occurred.
Synaptic cleft	Synaptic cleft refers to a microscopic gap between the terminal button of a neuron and the cell membrane of another neuron.
Sexual dysfunction	Sexual dysfunction or sexual malfunction is difficulty during any stage of the sexual act (which includes desire, arousal, orgasm, and resolution) that prevents the individual or couple from enjoying sexual activity.
Protein	A protein is a complex, high-molecular-weight organic compound that consists of amino acids joined by peptide bonds. It is essential to the structure and function of all living cells and viruses. Many are enzymes or subunits of enzymes.
Obsessive-compulsive disorder	Obsessive-compulsive disorder is an anxiety disorder manifested in a variety of forms, but is most commonly characterized by a subject's obsessive drive to perform a particular task or set of tasks, compulsions commonly termed rituals.
Guilt	Guilt describes many concepts related to a negative emotion or condition caused by actions which are believed to be, morally wrong. According to Freud, the avoidance of guilt is the basis for moral behavior.
Morphine	Morphine, the principal active agent in opium, is a powerful opioid analgesic drug. According to recent research, it may also be produced naturally by the human brain. Morphine is usually highly addictive, and tolerance and physical and psychological dependence develop quickly.
Opium	Opium is a narcotic analgesic drug which is obtained from the unripe seed pods of the opium poppy. Regular use, even for a few days, invariably leads to physical tolerance and dependence. Various degrees of psychological addiction can occur, though this is relatively rare when opioids are properly used..
Candace Pert	Candace Pert is the author of Molecules of Emotion and the neuroscientist who discovered the opiate receptor in the brain along with Solomon Snyder.
Opiates	A group of narcotics derived from the opium poppy that provide a euphoric rush and depress the nervous system are referred to as opiates.
Endorphin	An endorphin is an endogenous opioid biochemical compound. They are peptides produced by the pituitary gland and the hypothalamus, and they resemble the opiates in their abilities to produce analgesia and a sense of well-being. In other words, they work as "natural pain killers."
Epinephrine	Epinephrine is a hormone and a neurotransmitter. Epinephrine plays a central role in the short-term stress reaction—the physiological response to threatening or exciting conditions. It is secreted by the adrenal medulla. When released into the bloodstream, epinephrine binds to multiple receptors and has numerous effects throughout the body.
Sigmund Freud	Sigmund Freud was the founder of the psychoanalytic school, based on his theory that unconscious motives control much behavior, that particular kinds of unconscious thoughts and memories are the source of neurosis, and that neurosis could be treated through bringing these unconscious thoughts and memories to consciousness in psychoanalytic treatment.
Psychoanalyst	A psychoanalyst is a specially trained therapist who attempts to treat the individual by

	uncovering and revealing to the individual otherwise subconscious factors that are contributing to some undesirable behavor.
Local anesthetic	Local anesthetic drugs act mainly by inhibiting sodium influx through sodium-specific ion channels in the neuronal cell membrane, in particular the so-called voltage-gated sodium channels. When the influx of sodium is interrupted, an action potential cannot arise and signal conduction is thus inhibited.
Marijuana	Marijuana is the dried vegetable matter of the Cannabis sativa plant. It contains large concentrations of compounds that have medicinal and psychoactive effects when consumed, usually by smoking or eating.
Hallucination	A hallucination is a sensory perception experienced in the absence of an external stimulus, as distinct from an illusion, which is a misperception of an external stimulus. They may occur in any sensory modality - visual, auditory, olfactory, gustatory, tactile, or mixed.
Schizophrenia	Schizophrenia is characterized by persistent defects in the perception or expression of reality. A person suffering from untreated schizophrenia typically demonstrates grossly disorganized thinking, and may also experience delusions or auditory hallucinations
Perception	Perception is the process of acquiring, interpreting, selecting, and organizing sensory information.
Cannabis	The hemp plant from which marijuana, hashish, and THC are derived is the cannabis.
MDMA	MDMA, most commonly known today by the street name ecstasy, is a synthetic entactogen of the phenethylamine family whose primary effect is to stimulate the secretion of large amounts of serotonin as well as dopamine and noradrenaline in the brain, causing a general sense of openness, empathy, energy, euphoria, and well-being.
Asthma	Asthma is a complex disease characterized by bronchial hyperresponsiveness (BHR), inflammation, mucus production and intermittent airway obstruction.
Tetrahydroca-nabinol	The major active chemical in marijuana and hashish is a tetrahydrocannabinol. A number of studies indicate that it may provide medical benefits for cancer and AIDS patients by increasing appetite and decreasing nausea, and by blocking the spread of some cancer-causing Herpes simplex viruses.
Peyote	Peyote is a hallucinogen obtained from the root of the peyote cactus. The active ingredient is mescaline, an alkaloid.
Anandamide	Anandamide is a naturally occurring neurotransmitter found in the brain of animals, as well as other organs. Anandamide receptors were originally discovered as being sensitive to tetrahydrocannabinol, which is among the psychoactive cannabinoids found in marijuana.
Norepinephrine	Norepinephrine is released from the adrenal glands as a hormone into the blood, but it is also a neurotransmitter in the nervous system. As a stress hormone, it affects parts of the human brain where attention and impulsivity are controlled. Along with epinephrine, this compound effects the fight-or-flight response, activating the sympathetic nervous system to directly increase heart rate, release energy from fat, and increase muscle readiness.
Hallucinogenic	Certain drugs can affect the subjective qualities of perception, thought or emotion, resulting in altered interpretations of sensory input, alternate states of consciousness, or hallucinations. The term hallucinogenic is often broadly applied, especially in current scientific literature, to some or all of these substances.
Antipsychotic	The term antipsychotic is applied to a group of drugs used to treat psychosis.
Learning	Learning is a relatively permanent change in behavior that results from experience. Thus, to attribute a behavioral change to learning, the change must be relatively permanent and must

	result from experience.
Gene	A gene is an ultramicroscopic area of the chromosome. It is the smallest physical unit of the DNA molecule that carries a piece of hereditary information.
Delirium tremens	Delirium tremens refers to a condition characterized by sweating, restlessness, disorientation, and hallucinations. It occurs in some chronic alcohol users when there is a sudden decrease in usage.
Etiology	Etiology is the study of causation. The term is used in philosophy, physics and biology in reference to the causes of various phenomena. It is generally the study of why things occur, or even the reasons behind the way that things act.
Metabolism	Metabolism is the biochemical modification of chemical compounds in living organisms and cells.
Sensitization	Sensitization is a process whereby an organism is made more responsive to certain aspects of its environment. For example, increases in the effects of a drug as a result of repeated administration. Also known as reverse tolerance.
Basal ganglia	The basal ganglia are a group of nuclei in the brain associated with motor and learning functions.
Depressant	A depressant is a chemical agent that diminishes a body function or activity. The term is used in particular with regard to the central nervous system where these chemicals are known as neurotransmitters. They tend to act on the CNS by increasing the activity of a particular neurotransmitter known as gamma-aminobutyric acid (GABA).
Amphetamine psychosis	Amphetamine psychosis is a form of psychosis which can result from amphetamine or methamphetamine use. Typically it appears after large doses or chronic use, although in rare cases some people may become psychotic after relatively small doses.
Animal model	An animal model usually refers to a non-human animal with a disease that is similar to a human condition.
Chronic	Chronic refers to a relatively long duration, usually more than a few months.
Withdrawal symptoms	Withdrawal symptoms are physiological changes that occur when the use of a drug is stopped or dosage decreased.
Physical dependence	Physical dependence describes increased tolerance of a drug combined with a physical need of the drug to function. Abrupt cessation of the drug is typically associated with negative physical withdrawal symptoms. Physical dependence is distinguished from addiction. While addiction tends to describe psychological and behavioral attributes, physical dependence is defined primarily using physical and biological concepts.
Addiction	Addiction is an uncontrollable compulsion to repeat a behavior regardless of its consequences. Many drugs or behaviors can precipitate a pattern of conditions recognized as addiction, which include a craving for more of the drug or behavior, increased physiological tolerance to exposure, and withdrawal symptoms in the absence of the stimulus.
Habit	A habit is a response that has become completely separated from its eliciting stimulus. Early learning theorists used the term to describe S-R associations, however not all S-R associations become a habit, rather many are extinguished after reinforcement is withdrawn.
Substance abuse	Substance abuse refers to the overindulgence in and dependence on a stimulant, depressant, or other chemical substance, leading to effects that are detrimental to the individual's physical or mental health, or the welfare of others.
Substance dependence	Drug addiction, or substance dependence is the compulsive use of drugs, to the point where the user has no effective choice but to continue use.

Hypothesis	A specific statement about behavior or mental processes that is testable through research is a hypothesis.
Stages	Stages represent relatively discrete periods of time in which functioning is qualitatively different from functioning at other periods.
Sensation	Sensation is the first stage in the chain of biochemical and neurologic events that begins with the impinging of a stimulus upon the receptor cells of a sensory organ, which then leads to perception, the mental state that is reflected in statements like "I see a uniformly blue wall."
Mental Representation	Stage six of the sensorimotor substages, Mental representation, 18 months to 2 years, marks the beginnings of insight, or true creativity. This marks the passage into unique thought in Piaget's later three areas of development.
Associative learning	A metal connection made between two events is called associative learning.
Incentive	An incentive is what is expected once a behavior is performed. An incentive acts as a reinforcer.
Drug addiction	Drug addiction, or substance dependence is the compulsive use of drugs, to the point where the user has no effective choice but to continue use.
Opioid	An opioid is any agent that binds to opioid receptors, found principally in the central nervous system and gastrointestinal tract.
Sexually Transmitted Disease	Sexually transmitted disease is commonly transmitted between partners through some form of sexual activity, most commonly vaginal intercourse, oral sex, or anal sex.
Society	The social sciences use the term society to mean a group of people that form a semi-closed (or semi-open) social system, in which most interactions are with other individuals belonging to the group.
Dopamine hypothesis	The suggestion that schizophrenia may result from excess dopamine activity at certain synaptic sites is referred to as the dopamine hypothesis.
Ventral tegmental area	The ventral tegmental area is part of the midbrain, lying close to the substantia nigra and the red nucleus. It is rich in dopamine and serotonin neurons. It is considered to be part of the pleasure or reward system, one of the major sources of incentive and behavioral motivation.
Nucleus accumbens	A complex of neurons that is part of the brain's "pleasure pathway" responsible for the experience of reward is referred to as the nucleus accumbens.
Limbic system	The limbic system is a group of brain structures that are involved in various emotions such as aggression, fear, pleasure and also in the formation of memory. The limbic system affects the endocrine system and the autonomic nervous system. It consists of several subcortical structures located around the thalamus.
Midbrain	Located between the hindbrain and forebrain, a region in which many nerve-fiber systems ascend and descend to connect the higher and lower portions of the brain is referred to as midbrain. It is archipallian in origin, meaning its general architecture is shared with the most ancient of vertebrates. Dopamine produced in the subtantia nigra plays a role in motivation and habituation of species from humans to the most elementary animals such as insects.
Disinhibition	A temporary increase in the strength of an extinguished response caused by an unrelated stimulus event is referred to as disinhibition.

Reasoning | Reasoning is the act of using reason to derive a conclusion from certain premises. There are two main methods to reach a conclusion,deductive reasoning and inductive reasoning.

Time-out | Time-out is an operant conditioning punishment procedure in which, after unwanted behavior, the person is temporarily removed from a setting--in theory, a removal of stimuli resulting in an increase in the frequency of wanted behaviors; an example of negative reinforcement.

Alcohol myopia | Alcohol myopia refers to condition in which a person under the influence of alcohol ignores the long-range consequences of his or her behavior and responds only to the poorly thought-out immediate aspects of a situation.

Nearsightedness | A vision deficiency in which close objects are seen clearly but distant objects appear blurry is nearsightedness.

Fraternal twins | Fraternal twins usually occur when two fertilized eggs are implanted in the uterine wall at the same time. The two eggs form two zygotes, and these twins are therefore also known as dizygotic. Dizygotic twins are no more similar genetically than any siblings.

Identical twins | Identical twins occur when a single egg is fertilized to form one zygote (monozygotic) but the zygote then divides into two separate embryos. The two embryos develop into foetuses sharing the same womb. Monozygotic twins are genetically identical unless there has been a mutation in development, and they are almost always the same gender.

Concordance | Concordance as used in genetics means the presence of the same trait in both members of a pair of twins, or in sets of individuals. A twin study examines the concordance rates of twins having the same trait, especially a disease, which can help determine how much the disease is affected by genetics versus environment.

Alcoholism | A disorder that involves long-term, repeated, uncontrolled, compulsive, and excessive use of alcoholic beverages and that impairs the drinker's health and work and social relationships is called alcoholism.

Personality trait | According to the Diagnostic and Statistical Manual of the American Psychiatric Association, a personality trait is a "prominent aspect of personality that is exhibited in a wide range of important social and personal contexts. ...".

Trait | An enduring personality characteristic that tends to lead to certain behaviors is called a trait. The term trait also means a genetically inherited feature of an organism.

Glutamate | Glutamate is one of the 20 standard amino acids used by all organisms in their proteins. It is critical for proper cell function, but it is not an essential nutrient in humans because it can be manufactured from other compounds.

Thiamine | Thiamine, also known as vitamin B1, is a colorless compound with chemical formula $C_{12}H_{17}ClN_4OS$. Systemic thiamine deficiency can lead to myriad problems including neurodegeneration, wasting, and death. Well-known syndromes caused by lack of thiamine due to malnutrition or a diet high in thiaminase-rich foods include Wernicke-Korsakoff syndrome and beriberi, diseases also common in chronic abusers of alcohol.

Ecstasy | Ecstasy as an emotion is to be outside oneself, in a trancelike state in which an individual transcends ordinary consciousness and as a result has a heightened capacity for exceptional thought or experience. Ecstasy also refers to a relatively new hallucinogen that is chemically similar to mescaline and the amphetamines.

Nerve | A nerve is an enclosed, cable-like bundle of nerve fibers or axons, which includes the glia that ensheath the axons in myelin. Neurons are sometimes called nerve cells, though this term is technically imprecise since many neurons do not form nerves.

Neocortex | The neocortex is part of the cerebral cortex which covers most of the surface of the cerebral hemispheres including the frontal, parietal, occipital, and temporal lobes. Often seen as the

	hallmark of human intelligence, the role of this structure in the brain appears to be involved in conscious thought, spatial reasoning, and sensory perception.
Delusion	A false belief, not generally shared by others, and that cannot be changed despite strong evidence to the contrary is a delusion.
Psychosis	Psychosis is a generic term for mental states in which the components of rational thought and perception are severely impaired. Persons experiencing a psychosis may experience hallucinations, hold paranoid or delusional beliefs, demonstrate personality changes and exhibit disorganized thinking. This is usually accompanied by features such as a lack of insight into the unusual or bizarre nature of their behavior, difficulties with social interaction and impairments in carrying out the activities of daily living.
Phencyclidine	Phencyclidine is a dissociative psychedelic drug formerly used as an anaesthetic agent. Although the primary psychoactive effects of the drug only last hours, total elimination from the body is prolonged, typically extending over weeks.
Endocrine gland	An endocrine gland is one of a set of internal organs involved in the secretion of hormones into the blood. The other major type of gland is the exocrine glands, which secrete substances—usually digestive juices—into the digestive tract or onto the skin.
Testes	Testes are the male reproductive glands or gonads; this is where sperm develop and are stored.
Testosterone	Testosterone is a steroid hormone from the androgen group. It is the principal male sex hormone and the "original" anabolic steroid.
Homeostasis	Homeostasis is the property of an open system, especially living organisms, to regulate its internal environment so as to maintain a stable condition, by means of multiple dynamic equilibrium adjustments controlled by interrelated regulation mechanisms.
Nucleus	In neuroanatomy, a cluster of cell bodies of neurons within the central nervous system is a nucleus.
Gland	A gland is an organ in an animal's body that synthesizes a substance for release such as hormones, often into the bloodstream or into cavities inside the body or its outer surface.
Pancreas	The pancreas is a retroperitoneal organ that serves two functions: it produces juice containing digestive enzymes; and it produces several important hormones including insulin, glucagon, and several other hormones.
Diabetes	Diabetes is a medical disorder characterized by varying or persistent elevated blood sugar levels, especially after eating. All types of diabetes share similar symptoms and complications at advanced stages: dehydration and ketoacidosis, cardiovascular disease, chronic renal failure, retinal damage which can lead to blindness, nerve damage which can lead to erectile dysfunction, gangrene with risk of amputation of toes, feet, and even legs.
Progesterone	A female sex hormone that promotes growth of the sex organs and helps maintain pregnancy is called progesterone.
Estradiol	Estradiol is a sex hormone. Labelled the "female" hormone but also present in males it represents the major estrogen in humans. Critical for sexual functioning estradiol also supports bone growth.
Species	Species refers to a reproductively isolated breeding population.
Estrus	The estrus cycle refers to the recurring physiologic changes that are induced by reproductive hormones in most mammalian placental females (humans and great apes are the only mammals who undergo a menstrual cycle instead).
Stress hormones	Group of hormones including cortico steroids, that are involved in the body's physiological

	stress response are referred to as stress hormones.
Adrenaline	Adrenaline refers to a hormone produced by the adrenal medulla that stimulates sympathetic ANS activity and generally arouses people and heightens their emotional responsiveness.
Cortisol	Cortisol is a corticosteroid hormone that is involved in the response to stress; it increases blood pressure and blood sugar levels and suppresses the immune system. Synthetic cortisol, also known as hydrocortisone, is used as a drug mainly to fight allergies and inflammation.
Adaptation	Adaptation is a lowering of sensitivity to a stimulus following prolonged exposure to that stimulus. Behavioral adaptations are special ways a particular organism behaves to survive in its natural habitat.
Sympathetic	The sympathetic nervous system activates what is often termed the "fight or flight response". It is an automatic regulation system, that is, one that operates without the intervention of conscious thought.
Adrenal medulla	Composed mainly of hormone-producing chromaffin cells, the adrenal medulla is the principal site of the conversion of the amino acid tyrosine into the catecholamines epinephrine and norepinephrine (also called adrenaline and noradrenaline, respectively).
Adrenal glands	The adrenal glands sit atop the kidneys. They are chiefly responsible for regulating the stress response through the synthesis of corticosteroids and catecholamines, including cortisol and adrenalin.
Cell membrane	A component of every biological cell, the selectively permeable cell membrane is a thin and structured bilayer of phospholipid and protein molecules that envelopes the cell. It separates a cell's interior from its surroundings and controls what moves in and out.
Subjective experience	Subjective experience refers to reality as it is perceived and interpreted, not as it exists objectively.
Myopia	Myopia is a refractive defect of the eye in which light focuses in front of the retina. Those with myopia are often described as nearsighted or short-sighted in that they typically can see nearby objects clearly but distant objects appear blurred because the lens cannot flatten enough.
Psychotic behavior	A psychotic behavior is a severe psychological disorder characterized by hallucinations and loss of contact with reality.

Brain	The brain controls and coordinates most movement, behavior and homeostatic body functions such as heartbeat, blood pressure, fluid balance and body temperature. Functions of the brain are responsible for cognition, emotion, memory, motor learning and other sorts of learning. The brain is primarily made up of two types of cells: glia and neurons.
Learning	Learning is a relatively permanent change in behavior that results from experience. Thus, to attribute a behavioral change to learning, the change must be relatively permanent and must result from experience.
Analogy	An analogy is a comparison between two different things, in order to highlight some form of similarity. Analogy is the cognitive process of transferring information from a particular subject to another particular subject.
Gene	A gene is an ultramicroscopic area of the chromosome. It is the smallest physical unit of the DNA molecule that carries a piece of hereditary information.
Hormone	A hormone is a chemical messenger from one cell (or group of cells) to another. The best known are those produced by endocrine glands, but they are produced by nearly every organ system. The function of hormones is to serve as a signal to the target cells; the action of the hormone is determined by the pattern of secretion and the signal transduction of the receiving tissue.
Neuron	The neuron is the primary cell of the nervous system. They are found in the brain, the spinal cord, in the nerves and ganglia of the peripheral nervous system. It is a specialized cell that conducts impulses through the nervous system and contains three major parts: cell body, dendrites, and an axon. It can have many dendrites but only one axon.
Adolescence	The period of life bounded by puberty and the assumption of adult responsibilities is adolescence.
Frontal lobe	The frontal lobe comprises four major folds of cortical tissue: the precentral gyrus, superior gyrus and the middle gyrus of the frontal gyri, the inferior frontal gyrus. It has been found to play a part in impulse control, judgement, language, memory, motor function, problem solving, sexual behavior, socialization and spontaneity.
Maturation	The orderly unfolding of traits, as regulated by the genetic code is called maturation.
Embryo	A developed zygote that has a rudimentary heart, brain, and other organs is referred to as an embryo.
Species	Species refers to a reproductively isolated breeding population.
Stages	Stages represent relatively discrete periods of time in which functioning is qualitatively different from functioning at other periods.
Neural tube	The neural tube is the embryonal structure that gives rise to the brain and spinal cord.
Spinal cord	The spinal cord is a part of the vertebrate nervous system that is enclosed in and protected by the vertebral column (it passes through the spinal canal). It consists of nerve cells. The spinal cord carries sensory signals and motor innervation to most of the skeletal muscles in the body.
Hindbrain	The lowest level of the brain, consisting of the medulla, cerebellum, and pons is called the hindbrain.
Forebrain	The forebrain is the highest level of the brain. Key structures in the forebrain are the limbic system, thalamus, basal ganglia, hypothalamus, and cerebral cortex.
Midbrain	Located between the hindbrain and forebrain, a region in which many nerve-fiber systems ascend and descend to connect the higher and lower portions of the brain is referred to as midbrain. It is archipallian in origin, meaning its general architecture is shared with the

	most ancient of vertebrates. Dopamine produced in the subtantia nigra plays a role in motivation and habituation of species from humans to the most elementary animals such as insects.
Darwin	Darwin achieved lasting fame as originator of the theory of evolution through natural selection. His book Expression of Emotions in Man and Animals is generally considered the first text on comparative psychology.
Nervous system	The body's electrochemical communication circuitry, made up of billions of neurons is a nervous system.
Zygote	A zygote is a cell that is the result of fertilization. That is, two haploid cells—usually (but not always) a sperm cell from a male and an ovum from a female—merge into a single diploid cell called the zygote.
Depression	In everyday language depression refers to any downturn in mood, which may be relatively transitory and perhaps due to something trivial. This is differentiated from Clinical depression which is marked by symptoms that last two weeks or more and are so severe that they interfere with daily living.
Anatomy	Anatomy is the branch of biology that deals with the structure and organization of living things. It can be divided into animal anatomy (zootomy) and plant anatomy (phytonomy). Major branches of anatomy include comparative anatomy, histology, and human anatomy.
Cowan	Cowan regards working memory not as a separate system, but as a part of long-term memory. Representations in working memory are a subset of the representations in long-term memory.
Genitals	Genitals refers to the internal and external reproductive organs.
Scrotum	The scrotum is an external sack of skin that holds the testes.
Penis	The penis is the external male copulatory organ and the external male organ of urination. In humans, the penis is homologous to the female clitoris, as it develops from the same embryonic structure. It is capable of erection for use in copulation.
Testosterone	Testosterone is a steroid hormone from the androgen group. It is the principal male sex hormone and the "original" anabolic steroid.
Stem cells	Stem cells are primal undifferentiated cells which retain the ability to differentiate into other cell types. This ability allows them to act as a repair system for the body, replenishing other cells as long as the organism is alive.
Early adulthood	The developmental period beginning in the late teens or early twenties and lasting into the thirties is called early adulthood; characterized by an increasing self-awareness.
Human genome	The complete sequence or mapping of genes in the human body and their locations is the human genome. It is made up of 23 chromosome pairs with a total of about 3 billion DNA base pairs.
Chromosome	The DNA which carries genetic information in biological cells is normally packaged in the form of one or more large macromolecules called a chromosome. Humans normally have 46.
Protein	A protein is a complex, high-molecular-weight organic compound that consists of amino acids joined by peptide bonds. It is essential to the structure and function of all living cells and viruses. Many are enzymes or subunits of enzymes.
Cellular differentiation	Cellular differentiation is a concept from developmental biology describing the process by which cells acquire a "type". The morphology of a cell may change dramatically during differentiation, but the genetic material remains the same, with few exceptions.
Glial	Glial cells are non-neuronal cells that provide support and nutrition, maintain homeostasis, form myelin, and participate in signal transmission in the nervous system.

Neurogenesis | Neurogenesis literally means "birth of neurons". Neurogenesis is most prevalent during pre-natal development and is the process by which neurons are created to populate the growing brain.

Dendrite | A dendrite is a slender, typically branched projection of a nerve cell, or "neuron," which conducts the electrical stimulation received from other cells to the body or soma of the cell from which it projects. This stimulation arrives through synapses, which typically are located near the tips of the dendrites and away from the soma.

Synapse | A synapse is specialized junction through which cells of the nervous system signal to one another and to non-neuronal cells such as muscles or glands.

Myelin | Myelin is an electrically insulating fatty layer that surrounds the axons of many neurons, especially those in the peripheral nervous system. The main consequence of a myelin sheath is an increase in the speed at which impulses propagate along the myelinated fiber. The sheath continues to develop throughout childhood.

Axon | An axon, or "nerve fiber," is a long slender projection of a nerve cell, or "neuron," which conducts electrical impulses away from the neuron's cell body or soma.

Cerebral palsy | Cerebral palsy is a group of permanent disorders associated with developmental brain injuries that occur during fetal development, birth, or shortly after birth. It is characterized by a disruption of motor skills, with symptoms such as spasticity, paralysis, or seizures.

Placenta | A membrane that permits the exchange of nutrients and waste products between the mother and her developing child but does not allow the maternal and fetal bloodstreams to mix is the placenta.

Anoxia | Asphyxia is a condition of severely deficient supply of oxygen to the body. In the absence of remedial action it will very rapidly lead to unconsciousness and death. Anoxia means the pathological state in which tissues do not get (enough of) oxygen. Problems during childbirth can lead to the newborn experiencing asphyxia.

Umbilical cord | The umbilical cord is a tube that connects a developing embryo or fetus to its placenta. It contains major arteries and veins for the exchange of nutrient- and oxygen-rich blood between the embryo and placenta.

Seizure | A seizure is a temporary alteration in brain function expressed as a changed mental state, tonic or clonic movements and various other symptoms. They are due to temporary abnormal electrical activity of a group of brain cells.

Dendritic spine | A dendritic spine is a small membranous extrusion that protrudes from a dendrite and forms one half of a synapse. Changes in dendritic spine density underlie many brain functions, including motivation, learning, and memory. In particular, long-term memory is mediated in part by the growth of new dendritic spines to reinforce a particular neural pathway.

Cajal | Cajal, Nobel laureate, is considered to be the father of modern neuroscience. He postulated that the nervous system is made up of billions of separate nerve cells (neurons) and that nerve cells are polarized. He suggested that neurons communicate with each other via specialized junctions. This hypothesis became the basis of the neuron doctrine.

Cones | Cones are photoreceptors that transmit sensations of color, function in bright light, and used in visual acuity. Infants prior to months of age can only distinguish green and red indicating the cones are not fully developed; they can see all of the colors by 2 months of

Cerebral cortex | The cerebral cortex is the outermost layer of the cerebrum and has a grey color. It is made up of four lobes and it is involved in many complex brain functions including memory, perceptual awareness, "thinking", language and consciousness. The cerebral cortex receives sensory information from many different sensory organs eg: eyes, ears, etc. and processes the

	information.
Postnatal	Postnatal is the period beginning immediately after the birth of a child and extending for about six weeks. The period is also known as postpartum and, less commonly, puerperium.
Visual cortex	The visual cortex is the general term applied to both the primary visual cortex and the visual association area. Anatomically, the visual cortex occupies the entire occipital lobe, the inferior temporal lobe (IT), posterior parts of the parietal lobe, and a few small regions in the frontal lobe.
Synaptic density	Synaptic density is believed to be an important indication of the extent of connectivity between neurons.
Neural darwinism	Edelman's theory that groups of neurons are in constant competition with one another, each attempting to recruit adjacent neurons to their group and thus perform a particular function is called neural Darwinism.
Evolution	Commonly used to refer to gradual change, evolution is the change in the frequency of alleles within a population from one generation to the next. This change may be caused by different mechanisms, including natural selection, genetic drift, or changes in population (gene flow).
Trait	An enduring personality characteristic that tends to lead to certain behaviors is called a trait. The term trait also means a genetically inherited feature of an organism.
Guilford	Guilford observed that most individuals display a preference for either convergent or divergent thinking. Scientists and engineers typically prefer the former and artists and performers, the latter.
Apoptosis	Apoptosis is one of the main types of programmed cell death. As such, it is a process of deliberate suicide by an unwanted cell in a multicellular organism.
Neural network	A clusters of neurons that is interconnected to process information is referred to as a neural network.
Auditory system	The auditory system is the sensory system for the sense of hearing. On its path from the outside world to the forebrain, sound information is preserved and modified in many ways. It changes media twice, first from air to fluid, then from fluid to action potentials.
Perception	Perception is the process of acquiring, interpreting, selecting, and organizing sensory information.
Myelination	The process in which the nerve cells are covered and insulated with a layer of fat cells, which increases the speed at which information travels through the nervous system is referred to as myelination.
Oligodendrocyte	An Oligodendrocyte is a variety of neuroglia. Their main function is the myelination of nerve cells exclusively in the central nervous system of the higher vertebrates. A single oligodendrocyte can extend to about a dozen axons, wrapping around approximately 1mm of each and forming the myelin sheath.
Astrocyte	An astrocyte, also known as astroglia, is a characteristic star-shaped cell in the brain. They are the biggest cells found in brain tissue and outnumber the neurons ten to one. A commonly accepted function is to structure physically the brain. A second function is to provide neurons with nutrients such as glucose. They regulate the flow of nutrients provided by capillaries by forming the blood-brain barrier.
Problem solving	An attempt to find an appropriate way of attaining a goal when the goal is not readily available is called problem solving.
Pincer grasp	The use of the opposing thumb to grasp objects between the thumb and other fingers is referred to as the pincer grasp.

Motor cortex	Motor cortex refers to the section of cortex that lies in the frontal lobe, just across the central fissure from the sensory cortex. Neural impulses in the motor cortex are linked to muscular responses throughout the body.
Chronological age	Chronological age refers to the number of years that have elapsed since a person's birth.
Acquisition	Acquisition is the process of adapting to the environment, learning or becoming conditioned. In classical conditoning terms, it is the initial learning of the stimulus response link, which involves a neutral stimulus being associated with a unconditioned stimulus and becoming a conditioned stimulus.
Affect	A subjective feeling or emotional tone often accompanied by bodily expressions noticeable to others is called affect.
Individual differences	Individual differences psychology studies the ways in which individual people differ in their behavior. This is distinguished from other aspects of psychology in that although psychology is ostensibly a study of individuals, modern psychologists invariably study groups.
Cognitive development	The process by which a child's understanding of the world changes as a function of age and experience is called cognitive development.
Piaget	Piaget argued that young children's answers were qualitatively different than older children rather than quantitative. There are two major aspects to his theory: the process of coming to know and the stages we move through as we gradually acquire this ability.
Principle of conservation	The knowledge that the quantity of a substance remains the same even though its shape or other aspects of its physical appearance might change is the principle of conservation.
Concrete operational	According to Piaget, the period from 7 to 12 years of age, which is characterized by logical thought and a loss of egocentrism, is referred to as concrete operational stage. Conservation skills are formed - understanding that quantity, length or number of items is unrelated to the appearance of the object or items.
Formal operational	According to Piaget, the period from age 12 to adulthood, which is characterized by abstract thought is referred to as the formal operational stage.
Sensorimotor	The first of Piaget's stages is the Sensorimotor stage. This stage typically ranges from birth to 2 years. In this stage, children experience the world through their senses. During this stage, object permanence and stranger anxiety develop.
Reasoning	Reasoning is the act of using reason to derive a conclusion from certain premises. There are two main methods to reach a conclusion,deductive reasoning and inductive reasoning.
Senses	The senses are systems that consist of a sensory cell type that respond to a specific kind of physical energy, and that correspond to a defined region within the brain where the signals are received and interpreted.
Conservation	Conservation refers to the recognition that basic properties of substances such as weight and mass remain the same even when transformations merely alter their appearance.
Concrete operations	In Piaget's theory, the third major stage of cognitive development, in which children can decenter their perception, are less egocentric, and can think logically about concrete objects is called concrete operations.
Qualitative change	A qualitative change refers to a change in kind, structure, or organization, such as the change from nonverbal to verbal communication.
Object permanence	Object permanence is the term used to describe the awareness that objects continue to exist even when they are no longer visible. According to Piaget, object permance for the infant develops once the sensorimotor stage is complete.

Stranger anxiety	Stranger anxiety refers to an infant's fear and wariness of strangers; it typically appears in the second half of the first year of life.
Moral reasoning	Moral reasoning involves concepts of justice, whereas social conventional judgments are concepts of social organization.
Pretend play	According to Piaget and Smilansky, pretend play is the third cognitive level of play. It involves imaginary people or situations.
Egocentrism	The inability to distinguish between one's own perspective and someone else's is referred to as egocentrism.
Discrimination	In Learning theory, discrimination refers the ability to distinguish between a conditioned stimulus and other stimuli. It can be brought about by extensive training or differential reinforcement. In social terms, it is the denial of privileges to a person or a group on the basis of prejudice.
Basal ganglia	The basal ganglia are a group of nuclei in the brain associated with motor and learning functions.
Temporal lobe	The temporal lobe is part of the cerebrum. It lies at the side of the brain, beneath the lateral or Sylvian fissure. Adjacent areas in the superior, posterior and lateral parts of the temporal lobe are involved in high-level auditory processing.
Plasticity	The capacity for modification and change is referred to as plasticity.
Hebb	Hebb demonstrated that the rearing of rats in an enriched environment could alter neural development and that sensory - neural connections were shaped by experience. He is famous for developing the concept of neural nets. He also believed that learning early in life is of the incremental variety, whereas later it is cognitive, insightful, and more all-or-none.
Control group	A group that does not receive the treatment effect in an experiment is referred to as the control group or sometimes as the comparison group.
Intelligence test	An intelligence test is a standardized means of assessing a person's current mental ability, for example, the Stanford-Binet test and the Wechsler Adult Intelligence Scale.
Disuse	Disuse refers to theory that memory traces weaken when memories are not periodically used or retrieved.
Receptor	A sensory receptor is a structure that recognizes a stimulus in the internal or external environment of an organism. In response to stimuli the sensory receptor initiates sensory transduction by creating graded potentials or action potentials in the same cell or in an adjacent one.
Sperry	Sperry separated the corpus callosum, the area of the brain used to transfer signals between the right and left hemispheres, to treat epileptics. He then tested these patients with tasks that were known to be dependant on specific hemispheres of the brain and demonstrated that the two halves of the brain now had independent functions.
Retina	The retina is a thin layer of cells at the back of the eyeball. It is the part of the eye which converts light into nervous signals. The retina contains photoreceptor cells which receive the light; the resulting neural signals then undergo complex processing by other neurons of the retina, and are transformed into action potentials in retinal ganglion cells whose axons form the optic nerve.
Projection	Attributing one's own undesirable thoughts, impulses, traits, or behaviors to others is referred to as projection.
Infancy	The developmental period that extends from birth to 18 or 24 months is called infancy.

Critical period	A period of time when an innate response can be elicited by a particular stimulus is referred to as the critical period.
Abnormal behavior	An action, thought, or feeling that is harmful to the person or to others is called abnormal behavior.
Prenatal	Prenatal period refers to the time from conception to birth.
Imprinting	Imprinting describes any kind of critical period sensitive learning (learning occurring at a particular age or a particular life stage) that is rapid and apparently independent of the consequences of behavior.
Lorenz	Lorenz demonstrated how incubator-hatched geese would imprint on the first suitable moving stimulus they saw within what he called a "critical period" of about 36 hours shortly after hatching. Most famously, the goslings would imprint on Lorenz himself .
Sensitive period	A sensitive period is a developmental window in which a predisposed behavior is most likely to develop given appropriate stimulation. In linguistic theory, the period from about 18 months to puberty is when the brain is thought to be primed for learning language because of plasticity of the brain.
Attachment	Attachment is the tendency to seek closeness to another person and feel secure when that person is present.
Stimulus	A change in an environmental condition that elicits a response is a stimulus.
Harlow	Harlow and his famous wire and cloth surrogate mother monkey studies demonstrated that the need for affection created a stronger bond between mother and infant than did physical needs. He also found that the more discrimination problems the monkeys solved, the better they became at solving them.
Society	The social sciences use the term society to mean a group of people that form a semi-closed (or semi-open) social system, in which most interactions are with other individuals belonging to the group.
Standard deviation	In probability and statistics, the standard deviation is the most commonly used measure of statistical dispersion. Simply put, it measures how spread out the values in a data set are.
Stanford-Binet	Terman released the "Stanford Revision of the Binet-Simon Scale" or the Stanford-Binet for short. Using validation experiments, he removed several of the Binet-Simon test items and added new ones. In 1985 it was revised to analyze an individual's responses in four content areas: verbal reasoning, quantitative reasoning, abstract reasoning, and short-term memory.
Brain-imaging	Brain-imaging is a fairly recent discipline within medicine and neuroscience. Brain imaging falls into two broad categories -- structural imaging and functional imaging.
Androgen	Androgen is the generic term for any natural or synthetic compound, usually a steroid hormone, that stimulates or controls the development and maintenance of masculine characteristics in vertebrates by binding to androgen receptors.
Deprivation	Deprivation, is the loss or withholding of normal stimulation, nutrition, comfort, love, and so forth; a condition of lacking. The level of stimulation is less than what is required.
Social skills	Social skills are skills used to interact and communicate with others to assist status in the social structure and other motivations.
Population	Population refers to all members of a well-defined group of organisms, events, or things.
Testes	Testes are the male reproductive glands or gonads; this is where sperm develop and are stored.
Estrogen	Estrogen is a group of steroid compounds that function as the primary female sex hormone.

	They are produced primarily by developing follicles in the ovaries, the corpus luteum and the placenta.
Chronic	Chronic refers to a relatively long duration, usually more than a few months.
Neuroscience	A field that combines the work of psychologists, biologists, biochemists, medical researchers, and others in the study of the structure and function of the nervous system is neuroscience.
Lesion	A lesion is a non-specific term referring to abnormal tissue in the body. It can be caused by any disease process including trauma (physical, chemical, electrical), infection, neoplasm, metabolic and autoimmune.
Spina bifida	Spina bifida are birth defects caused by an incomplete closure of one or more vertebral arches of the spine, resulting in malformations of the spinal cord. Spina bifida results in varying degrees of paralysis, absence of skin sensation, incontinence, and spine and limb problems depending on the severity and location of the lesion damage on the spine.
Schizophrenia	Schizophrenia is characterized by persistent defects in the perception or expression of reality. A person suffering from untreated schizophrenia typically demonstrates grossly disorganized thinking, and may also experience delusions or auditory hallucinations
Delusion	A false belief, not generally shared by others, and that cannot be changed despite strong evidence to the contrary is a delusion.
Affective	Affective is the way people react emotionally, their ability to feel another living thing's pain or joy.
Kraepelin	Kraepelin postulated that there is a specific brain or other biological pathology underlying each of the major psychiatric disorders. Just as his laboratory discovered the pathologic basis of what is now known as Alzheimers disease, Kraepelin was confident that it would someday be possible to identify the pathologic basis of each of the major psychiatric disorders.
Excitatory synapse	An excitatory synapse is where the neurotransmitter increases the likelihood that an action potential will occur, or increases the rate at which they are already occurring, in the neuron on which it acts.
Prefrontal cortex	The prefrontal cortex is the anterior part of the frontal lobes of the brain, lying in front of the motor and associative areas. It has been implicated in planning complex cognitive behaviors, personality expression and moderating correct social behavior. The prefrontal cortex continues to develop until around age 6.
Mental retardation	Mental retardation refers to having significantly below-average intellectual functioning and limitations in at least two areas of adaptive functioning. Many categorize retardation as mild, moderate, severe, or profound.
Late adolescence	Late adolescence refers to approximately the latter half of the second decade of life. Career interests, dating, and identity exploration are often more pronounced in late adolescence than in early adolescence.
Syndrome	The term syndrome is the association of several clinically recognizable features, signs, symptoms, phenomena or characteristics which often occur together, so that the presence of one feature indicates the presence of the others.
Measles	Measles, also known as rubeola, is a common disease caused by a virus of the genus Morbillivirus. Complications with measles are relatively common, ranging from relatively common and less serious diarrhea, to pneumonia and encephalitis. Complications are usually more severe amongst infants and adults who catch the virus.

Golgi	Golgi discovered a method of staining nervous tissue which would stain a limited number of cells at random, in their entirety. This enabled him to view the paths of nerve cells in the brain for the first time. He called his discovery the black reaction. It is now known universally as the Golgi stain.
Phenotype	The phenotype of an individual organism is either its total physical appearance and constitution, or a specific manifestation of a trait, such as size or eye color, that varies between individuals. Phenotype is determined to some extent by genotype, or by the identity of the alleles that an individual carries at one or more positions on the chromosomes.
Physiology	The study of the functions and activities of living cells, tissues, and organs and of the physical and chemical phenomena involved is referred to as physiology.
Correlation	A statistical technique for determining the degree of association between two or more variables is referred to as correlation.
Trauma	Trauma refers to a severe physical injury or wound to the body caused by an external force, or a psychological shock having a lasting effect on mental life.
Hypothesis	A specific statement about behavior or mental processes that is testable through research is a hypothesis.

Blind spot	In anatomy, the blind spot is the region of the retina where the optic nerve and blood vessels pass through to connect to the back of the eye. Since there are no light receptors there, a part of the field of vision is not perceived.
Sensation	Sensation is the first stage in the chain of biochemical and neurologic events that begins with the impinging of a stimulus upon the receptor cells of a sensory organ, which then leads to perception, the mental state that is reflected in statements like "I see a uniformly blue wall."
Occipital lobe	The occipital lobe is the smallest of four true lobes in the human brain. Located in the rearmost portion of the skull, the occipital lobe is part of the forebrain structure. It is the visual processing center.
Neurologist	A physician who studies the nervous system, especially its structure, functions, and abnormalities is referred to as neurologist.
Migraine	Migraine is a form of headache, usually very intense and disabling. It is a neurologic disease.
Brain	The brain controls and coordinates most movement, behavior and homeostatic body functions such as heartbeat, blood pressure, fluid balance and body temperature. Functions of the brain are responsible for cognition, emotion, memory, motor learning and other sorts of learning. The brain is primarily made up of two types of cells: glia and neurons.
Lashley	Lashley failed to find a single biological locus of memory suggesting to him that memories were not localized to one part of the brain, but were widely distributed throughout the cortex.
Population	Population refers to all members of a well-defined group of organisms, events, or things.
Tactile	Pertaining to the sense of touch is referred to as tactile.
Scotoma	The Scotoma is an area of loss of visual acuity surrounded by a field of normal or relatively well-preserved vision. Every normal eye has a scotoma in its field of vision, usually termed its blind spot. Others may occur from damage.
Perception	Perception is the process of acquiring, interpreting, selecting, and organizing sensory information.
Limited capacity	Limited capacity refers to the concept that one's information processing ability is restricted. Metaphors for capacity include mental space, mental energy or effort, and time.
Senses	The senses are systems that consist of a sensory cell type that respond to a specific kind of physical energy, and that correspond to a defined region within the brain where the signals are received and interpreted.
Color blindness	Color blindness in humans is the inability to perceive differences between some or all colors that other people can distinguish. It is most often of genetic nature, but may also occur because of eye, nerve, or brain damage, or due to exposure to certain chemicals.
Receptor	A sensory receptor is a structure that recognizes a stimulus in the internal or external environment of an organism. In response to stimuli the sensory receptor initiates sensory transduction by creating graded potentials or action potentials in the same cell or in an adjacent one.
Gene	A gene is an ultramicroscopic area of the chromosome. It is the smallest physical unit of the DNA molecule that carries a piece of hereditary information.
Ophthalmoscope	In 1851, Helmholtz invented the ophthalmoscope, an instrument which can be used to look into the human eye.

Retina	The retina is a thin layer of cells at the back of the eyeball. It is the part of the eye which converts light into nervous signals. The retina contains photoreceptor cells which receive the light; the resulting neural signals then undergo complex processing by other neurons of the retina, and are transformed into action potentials in retinal ganglion cells whose axons form the optic nerve.
Sclera	The sclera is the white outer coating of the eye made of tough fibrin connective tissue which gives the eye its shape and helps to protect the delicate inner parts.
Cornea	The cornea is the transparent front part of the eye that covers the iris, pupil, and anterior chamber and provides most of an eye's optical power. Together with the lens, the cornea refracts light and consequently helps the eye to focus.
Fovea	The fovea, a part of the eye, is a spot located in the center of the macula. The fovea is responsible for sharp central vision, which is necessary in humans for reading, watching television or movies, driving, and any activity where visual detail is of primary importance.
Nerve	A nerve is an enclosed, cable-like bundle of nerve fibers or axons, which includes the glia that ensheath the axons in myelin. Neurons are sometimes called nerve cells, though this term is technically imprecise since many neurons do not form nerves.
Optic nerve	The optic nerve is the nerve that transmits visual information from the retina to the brain. The optic nerve is composed of retinal ganglion cell axons and support cells.
Ganglion cell	A ganglion cell is a type of neuron located in the retina of the eye that receives visual information from photoreceptors via various intermediate cells such as bipolar cells, amacrine cells, and horizontal cells. Retinal ganglion cells' axons are myelinated.
Axon	An axon, or "nerve fiber," is a long slender projection of a nerve cell, or "neuron," which conducts electrical impulses away from the neuron's cell body or soma.
Photoreceptor	Photoreceptor cells are contained in the retina and are responsible for transducing, or converting, light into signals that can be ultimately transmitted to the brain via the optic nerve. A photoreceptor cell contains a membranous protein called an opsin which contains a pigment molecule called retinal.
Neuron	The neuron is the primary cell of the nervous system. They are found in the brain, the spinal cord, in the nerves and ganglia of the peripheral nervous system. It is a specialized cell that conducts impulses through the nervous system and contains three major parts: cell body, dendrites, and an axon. It can have many dendrites but only one axon.
Presbyopia	Presbyopia is the eye's diminished power of accommodation that occurs with aging. Presbyopia is not a disease as such, but a condition that affects everyone at a certain age.
Cerebrospinal fluid	A solution that fills the hollow cavities of the brain and circulates around the brain and spinal cord is called cerebrospinal fluid.
Tumor	A tumor is an abnormal growth that when located in the brain can either be malignant and directly destroy brain tissue, or be benign and disrupt functioning by increasing intracranial pressure.
Depression	In everyday language depression refers to any downturn in mood, which may be relatively transitory and perhaps due to something trivial. This is differentiated from Clinical depression which is marked by symptoms that last two weeks or more and are so severe that they interfere with daily living.
Cones	Cones are photoreceptors that transmit sensations of color, function in bright light, and used in visual acuity. Infants prior to months of age can only distinguish green and red indicating the cones are not fully developed; they can see all of the colors by 2 months of

Rods	Rods are cylindrical shaped photoreceptors that are sensitive to the intensity of light. Rods require less light to function than cone cells, and therefore are the primary source of visual information at night.
Receptive field	The receptive field of a sensory neuron is a region of sensitivity in which the presence of a stimulus will alter the firing of that neuron.
Visual acuity	Visual acuity is the eye's ability to detect fine details and is the quantitative measure of the eye's ability to see an in-focus image at a certain distance.
X chromosome	The sex chromosomes are one of the 23 pairs of human chromosomes. Each person normally has one pair of sex chromosomes in each cell. Females have two X chromosomes, while males have one X and one Y chromosome. The X chromosome carries hundreds of genes but few, if any, of these have anything to do directly with sex determination.
Horizontal cells	Horizontal cells are the laterally interconnecting neurons in the outer plexiform layer of the retina.
Amacrine cells	Amacrine cells are interneurons in the retina which operate at the Inner Plexiform Layer (IPL), the second synaptic retinal layer where bipolar cells and ganglion cells synapse. Functionally, they are responsible for complex processing of the retinal image, specifically adjusting image brightness and, by integrating sequential activation of neurons, detecting motion.
Bipolar cell	As a part of the retina, the bipolar cell exists between photoreceptors and ganglion cells. Bipolar cells are so-named as they have a central body from which two sets of processes arise. At one end, they form synapses with either a single cone cell, or a number of rod cells. At the other end, they form synapses with ganglion cells, which fire action potentials along the optic nerve (cranial nerve II). They effectively transfer information from rods and cones to ganglion cells.
Action potential	The electrical impulse that provides the basis for the conduction of a neural impulse along an axon of a neuron is the action potential. When a biological cell or patch of membrane undergoes an action potential, or electrical excitation, the polarity of the transmembrane voltage swings rapidly from negative to positive and back.
Thalamus	An area near the center of the brain involved in the relay of sensory information to the cortex and in the functions of sleep and attention is the thalamus.
Lateral geniculate nucleus	The lateral geniculate nucleus of the thalamus is a part of the brain, which is the primary processor of visual information, received from the retina, in the CNS.
Visual cortex	The visual cortex is the general term applied to both the primary visual cortex and the visual association area. Anatomically, the visual cortex occupies the entire occipital lobe, the inferior temporal lobe (IT), posterior parts of the parietal lobe, and a few small regions in the frontal lobe.
Projection	Attributing one's own undesirable thoughts, impulses, traits, or behaviors to others is referred to as projection.
Primary visual cortex	Primary visual cortex refers to the area at the rear of the occipital lobes where vision registers in the cerebral cortex.
Optic chiasm	Optic chiasm refers to the point at which the optic nerves from the inside half of each eye cross over and then project to the opposite half of the brain.
Right hemisphere	The brain is divided into left and right cerebral hemispheres. The right hemisphere of the cortex controls the left side of the body.

Left hemisphere	The left hemisphere of the cortex controls the right side of the body, coordinates complex movements, and, in 95% of people, controls the production of speech and written language.
Striate cortex	The functionally defined primary visual cortex is approximately equivalent to the anatomically defined striate cortex located in the occipital lobe.
Parietal lobe	The parietal lobe is positioned above (superior to) the occipital lobe and behind (posterior to) the frontal lobe. It plays important roles in integrating sensory information from various senses, and in the manipulation of objects.
Temporal lobe	The temporal lobe is part of the cerebrum. It lies at the side of the brain, beneath the lateral or Sylvian fissure. Adjacent areas in the superior, posterior and lateral parts of the temporal lobe are involved in high-level auditory processing.
Ventral stream	The primate visual system consists of about thirty areas of the cerebral cortex called the visual cortex. The visual cortex is divided into the ventral stream and the dorsal stream. The ventral stream is associated with object recognition and form representation.
Evolution	Commonly used to refer to gradual change, evolution is the change in the frequency of alleles within a population from one generation to the next. This change may be caused by different mechanisms, including natural selection, genetic drift, or changes in population (gene flow).
Stimulus	A change in an environmental condition that elicits a response is a stimulus.
Dorsal stream	The dorsal stream is a pathway for visual information which flows through the visual cortex, the part of the brain which provides visual processing. It is involved in spatial awareness: recognizing where objects are in space.
Society	The social sciences use the term society to mean a group of people that form a semi-closed (or semi-open) social system, in which most interactions are with other individuals belonging to the group.
Extrastriate cortex	The extrastriate cortex is the locus of mid-level vision. Neurons in the extrastriate cortex generally respond to visual stimuli within their receptive fields. These responses are modulated by extraretinal effects, like attention, working memory, and reward expectation.
Metabolism	Metabolism is the biochemical modification of chemical compounds in living organisms and cells.
Enzyme	An enzyme is a protein that catalyzes, or speeds up, a chemical reaction. Enzymes are essential to sustain life because most chemical reactions in biological cells would occur too slowly, or would lead to different products, without enzymes.
Heterogeneous	A heterogeneous compound, mixture, or other such object is one that consists of many different items, which are often not easily sorted or separated, though they are clearly distinct.
Staining	Staining is a biochemical technique of adding a class-specific (DNA, proteins, lipids, carbohydrates) dye to a substrate to qualify or quantify the presence of a specific compound. They are frequently used to highlight structures in tissues for viewing, often with the aid of different microscopes.
Motion perception	Motion perception is the process of inferring the "true" velocity and direction of motion in a visual scene given some visual input.
Mitochondria	Mitochondria are sometimes described as "cellular power plants", because their primary function is to convert organic materials into energy in the form of ATP.
Nucleus	In neuroanatomy, a cluster of cell bodies of neurons within the central nervous system is a nucleus.

Midbrain	Located between the hindbrain and forebrain, a region in which many nerve-fiber systems ascend and descend to connect the higher and lower portions of the brain is referred to as midbrain. It is archipallian in origin, meaning its general architecture is shared with the most ancient of vertebrates. Dopamine produced in the subtantia nigra plays a role in motivation and habituation of species from humans to the most elementary animals such as insects.
Coding	In senation, coding is the process by which information about the quality and quantity of a stimulus is preserved in the pattern of action potentials sent through sensory neurons to the central nervous system.
Physiology	The study of the functions and activities of living cells, tissues, and organs and of the physical and chemical phenomena involved is referred to as physiology.
Synapse	A synapse is specialized junction through which cells of the nervous system signal to one another and to non-neuronal cells such as muscles or glands.
Frontal lobe	The frontal lobe comprises four major folds of cortical tissue: the precentral gyrus, superior gyrus and the middle gyrus of the frontal gyri, the inferior frontal gyrus. It has been found to play a part in impulse control, judgement, language, memory, motor function, problem solving, sexual behavior, socialization and spontaneity.
Corpus callosum	The corpus callosum is the largest white matter structure in the brain. It consists of mostly of contralateral axon projections. The corpus callosum connects the left and right cerebral hemispheres. Most communication between regions in different halves of the brain are carried over the corpus callosum.
Microelectrode	An electrical wire so small that it can be used either to monitor the electrical activity of a single neuron or to stimulate activity within it is a microelectrode.
Electrode	Any device used to electrically stimulate nerve tissue or to record its activity is an electrode.
Nervous system	The body's electrochemical communication circuitry, made up of billions of neurons is a nervous system.
Baseline	Measure of a particular behavior or process taken before the introduction of the independent variable or treatment is called the baseline.
Off-center	The ganglion off-center cell in the retina is stimulated by activation of the surround and inhibited by stimulation of the center.
On-center	The ganglion on-center cell is stimulated when the center of its receptive field is exposed to light, and is inhibited when the surround is exposed to light.
Neuroscience	A field that combines the work of psychologists, biologists, biochemists, medical researchers, and others in the study of the structure and function of the nervous system is neuroscience.
Affect	A subjective feeling or emotional tone often accompanied by bodily expressions noticeable to others is called affect.
Simple cell	A neuron in the striate cortex that is maximally sensitive to the position and orientation of edges in the receptive field is called a simple cell.
Trichromatic theory	The trichromatic theory was postulated by Young and later by Helmholtz. They demonstrated that most colors can be matched by superimposing three separate light sources known as primaries; a process known as additive mixing. The Young-Helmholtz theory of color vision was built around the assumption of there being three classes of receptors.
Opponent-	The opponent-process theory is a color theory that states that the human visual system

process theory	interprets information about color by processing signals from cones in an antagonistic manner.
Afterimage	An afterimage is an optical illusion that occurs after looking away from a direct gaze at an image. This is closely related to the phenomenon called the persistence of vision, which is used in animation and cinema.
Hering	Hering disagreed with the trichromatic theory of color perception which held that the human eye perceived all colors in terms of three primary colors. (Red, Green, Blue). He believed that the visual system worked based on a system of color opponency, a system of six primary colors.
Color constancy	Color constancy is an example of subjective constancy and a feature of the human color-perception system which ensures that the perceived color of objects remains relatively constant under varying illumination conditions in spite of physical change.
Lesion	A lesion is a non-specific term referring to abnormal tissue in the body. It can be caused by any disease process including trauma (physical, chemical, electrical), infection, neoplasm, metabolic and autoimmune.
Anesthesia	Anesthesia is the process of blocking the perception of pain and other sensations. This allows patients to undergo surgery and other procedures without the distress and pain they would otherwise experience.
Agnosia	Agnosia is a loss of ability to recognize objects, persons, sounds, shapes or smells while the specific sense is not defective nor is there any significant memory loss. It is usually associated with brain injury or neurological illness, particularly after damage to the temporal lobe.
Anoxia	Asphyxia is a condition of severely deficient supply of oxygen to the body. In the absence of remedial action it will very rapidly lead to unconsciousness and death. Anoxia means the pathological state in which tissues do not get (enough of) oxygen. Problems during childbirth can lead to the newborn experiencing asphyxia.
Lungs	The lungs are the essential organs of respiration. Its principal function is to transport oxygen from the atmosphere into the bloodstream, and excrete carbon dioxide from the bloodstream into the atmosphere.
Cerebral cortex	The cerebral cortex is the outermost layer of the cerebrum and has a grey color. It is made up of four lobes and it is involved in many complex brain functions including memory, perceptual awareness, "thinking", language and consciousness. The cerebral cortex receives sensory information from many different sensory organs eg: eyes, ears, etc. and processes the information.
Hippocampus	The hippocampus is a part of the brain located inside the temporal lobe. It forms a part of the limbic system and plays a part in memory and navigation.
Cerebellum	The cerebellum is located in the inferior posterior portion of the head (the hindbrain), directly dorsal to the brainstem and pons, inferior to the occipital lobe. The cerebellum is a region of the brain that plays an important role in the integration of sensory perception and fine motor output.
Variable	A variable refers to a measurable factor, characteristic, or attribute of an individual or a system.
Ataxia	Ataxia is unsteady and clumsy motion of the limbs or trunk due to a failure of the fine coordination of muscle movements.
Illusion	An illusion is a distortion of a sensory perception.

Hue	A hue refers to the gradation of color within the optical spectrum, or visible spectrum, of light. Hue may also refer to a particular color within this spectrum, as defined by its dominant wavelength, or the central tendency of its combined wavelengths.
Case study	A carefully drawn biography that may be obtained through interviews, questionnaires, and psychological tests is called a case study.
Blindsight	Blindsight is residual visual sensitivity independent of any subjective experience of visual perception. A person experiencing blindsight, if not consciously able to see, is still influenced by objects in his field of vision.
Neuropsychology	Neuropsychology is a branch of psychology that aims to understand how the structure and function of the brain relates to specific psychological processes.

Brain	The brain controls and coordinates most movement, behavior and homeostatic body functions such as heartbeat, blood pressure, fluid balance and body temperature. Functions of the brain are responsible for cognition, emotion, memory, motor learning and other sorts of learning. The brain is primarily made up of two types of cells: glia and neurons.
Temporal lobe	The temporal lobe is part of the cerebrum. It lies at the side of the brain, beneath the lateral or Sylvian fissure. Adjacent areas in the superior, posterior and lateral parts of the temporal lobe are involved in high-level auditory processing.
Displacement	An unconscious defense mechanism in which the individual directs aggressive or sexual feelings away from the primary object to someone or something safe is referred to as displacement. Displacement in linguistics is simply the ability to talk about things not present.
Amplitude	Amplitude is a nonnegative scalar measure of a wave's magnitude of oscillation.
Auditory system	The auditory system is the sensory system for the sense of hearing. On its path from the outside world to the forebrain, sound information is preserved and modified in many ways. It changes media twice, first from air to fluid, then from fluid to action potentials.
Perception	Perception is the process of acquiring, interpreting, selecting, and organizing sensory information.
Pitch	Pitch is the psychological interpretation of a sound or musical tone corresponding to its physical frequency
Loudness	Loudness is the quality of a sound that is the primary psychological correlate of physical intensity. Loudness is often approximated by a power function with an exponent of 0.6 when plotted vs. sound pressure or 0.3 when plotted vs. sound intensity.
Timbre	The distinctive quality of a sound that distinguishes it from other sounds of the same pitch and loudness is called timbre.
Species	Species refers to a reproductively isolated breeding population.
Nervous system	The body's electrochemical communication circuitry, made up of billions of neurons is a nervous system.
Threshold	In general, a threshold is a fixed location or value where an abrupt change is observed. In the sensory modalities, it is the minimum amount of stimulus energy necessary to elicit a sensory response.
Fundamental frequency	The fundamental frequency of a periodic signal is the inverse of the pitch period length. The pitch period is, in turn, the smallest repeating unit of a signal.
Receptor	A sensory receptor is a structure that recognizes a stimulus in the internal or external environment of an organism. In response to stimuli the sensory receptor initiates sensory transduction by creating graded potentials or action potentials in the same cell or in an adjacent one.
Learning	Learning is a relatively permanent change in behavior that results from experience. Thus, to attribute a behavioral change to learning, the change must be relatively permanent and must result from experience.
Prosody	Prosody consists of the distinctive variations of stress, tone, and timing in spoken language. How pitch changes from word to word, the speed of speech, the loudness of speech, and the duration of pauses all contribute to prosody.
Anatomy	Anatomy is the branch of biology that deals with the structure and organization of living things. It can be divided into animal anatomy (zootomy) and plant anatomy (phytonomy). Major branches of anatomy include comparative anatomy, histology, and human anatomy.

Inner ear	The inner ear consists of the oval window, cochlea, and basilar membrane.
Outer ear	Outer ear consists of the pinna and the external auditory canal.
Pinna	The pinna is the visible part of the ear that resides outside of the head. It acts as a funnel, amplifying the sound and directing it to the ear canal. While reflecting from the pinna, sound also goes through a filtering process which adds directional information to the sound.
Eardrum	The tympanum or tympanic membrane, colloquially known as the eardrum, is a thin membrane that separates the outer ear from the middle ear. Its function is to transmit sound from the air to the ossicles inside the middle ear. The malleus bone connects the eardrum to the other ossicles.
Middle ear	The middle ear consists of the eardrum, hammer, anvil, and stirrup.
Ossicles	The ossicles are the three smallest bones in the human body. They are contained within the middle ear space and serve to transmit sounds from the air to the fluid filled labyrinth (cochlea).
Oval window	The oval window is a membrane-covered opening which leads from the middle ear to the vestibule of the inner ear.
Cochlea	The Cochlea is the bony tube that contains the basilar membrane and the organ of Corti. The cochlea consists of three fluid-filled chambers - scala tympani and scala vestibuli and scala media.
Hair cells	Hair cells are the sensory cells of both the auditory system and the vestibular system. The auditory hair cells are located within the organ of Corti on a thin basilar membrane in the cochlea of the inner ear.
Basilar membrane	The basilar membrane within the cochlea of the inner ear is the part of the auditory system that decomposes incoming auditory signals into their frequency components. This allows higher neural processing of sound information to focus on the frequency spectrum of input, rather than just the time domain waveform.
Semicircular canals	The semicircular canals are three half-circular, interconnected tubes located inside each ear that are the equivalent of three gyroscopes located in three planes perpendicular (at right angles) to each other.
Nerve	A nerve is an enclosed, cable-like bundle of nerve fibers or axons, which includes the glia that ensheath the axons in myelin. Neurons are sometimes called nerve cells, though this term is technically imprecise since many neurons do not form nerves.
Auditory nerve	The vestibulocochlear nerve is the eighth of twelve cranial nerves, and also known as the auditory nerve. It is the nerve along which the sensory cells (the hair cells) of the inner ear transmit information to the brain. It consists of the cochlear nerve, carrying information about hearing, and the vestibular nerve, carrying information about balance.
Axon	An axon, or "nerve fiber," is a long slender projection of a nerve cell, or "neuron," which conducts electrical impulses away from the neuron's cell body or soma.
Organ of Corti	The Organ of Corti is the hearing organ of the inner ear. It contains receptors that respond to vibrations in the basilar membrane which are caused by sound.
Helmholtz	Helmholtz a pioneer of the new science of psychology, was a rigorous experimental physiologist and philospher. He gave us the distinction betwen sensation and peception and is well known for his theories of color perception and hearing.
Traveling wave	The pattern of vibration in the cochlear fluid, which varies as a function of the amplitude and frequency of an airborne sound is referred to as a traveling wave.

Transduction	Transduction in the nervous system typically refers to synaptic events wherein an electrical signal, known as an action potential, is converted into a chemical one via the release of neurotransmitters. Conversely, in sensory transduction a chemical or physical stimulus is transduced by sensory receptors into an electrical signal.
Action potential	The electrical impulse that provides the basis for the conduction of a neural impulse along an axon of a neuron is the action potential. When a biological cell or patch of membrane undergoes an action potential, or electrical excitation, the polarity of the transmembrane voltage swings rapidly from negative to positive and back.
Ion	An ion is an atom or group of atoms with a net electric charge. The energy required to detach an electron in its lowest energy state from an atom or molecule of a gas with less net electric charge is called the ionization potential, or ionization energy.
Polarization	Polarization is the process of preparing a neuron for firing by creating an internal negative charge in relation to the body fluid outside the cell membrane.
Depolarization	Depolarization is any change that decreases the absolute value of a cell's membrane potential. Thus, changes in membrane voltage in which the membrane potential becomes less positive or less negative are both depolarizations.
Dendrite	A dendrite is a slender, typically branched projection of a nerve cell, or "neuron," which conducts the electrical stimulation received from other cells to the body or soma of the cell from which it projects. This stimulation arrives through synapses, which typically are located near the tips of the dendrites and away from the soma.
Hyperpolariz-tion	Hyperpolarization is any change in a cell's membrane potential that makes it more polarized. That is, hyperpolarization is an increase in the absolute value of a cell's membrane potential.
Cranial nerve	A Cranial nerve emerges from the brainstem instead of the spinal cord.
Synapse	A synapse is specialized junction through which cells of the nervous system signal to one another and to non-neuronal cells such as muscles or glands.
Projection	Attributing one's own undesirable thoughts, impulses, traits, or behaviors to others is referred to as projection.
Nucleus	In neuroanatomy, a cluster of cell bodies of neurons within the central nervous system is a nucleus.
Thalamus	An area near the center of the brain involved in the relay of sensory information to the cortex and in the functions of sleep and attention is the thalamus.
Primary auditory cortex	The primary auditory cortex is responsible for processing of auditory information. It is located in the temporal lobe; the posterior half of the superior temporal gyrus and also dives into the lateral sulcus as the transverse temporal gyri.
Hindbrain	The lowest level of the brain, consisting of the medulla, cerebellum, and pons is called the hindbrain.
Midbrain	Located between the hindbrain and forebrain, a region in which many nerve-fiber systems ascend and descend to connect the higher and lower portions of the brain is referred to as midbrain. It is archipallian in origin, meaning its general architecture is shared with the most ancient of vertebrates. Dopamine produced in the subtantia nigra plays a role in motivation and habituation of species from humans to the most elementary animals such as insects.
Gyrus	A gyrus is a ridge on the cerebral cortex. It is generally surrounded by one or more sulci.
Left hemisphere	The left hemisphere of the cortex controls the right side of the body, coordinates complex

	movements, and, in 95% of people, controls the production of speech and written language.
Lateralization	Lateralization refers to the dominance of one hemisphere of the brain for specific functions.
Neuron	The neuron is the primary cell of the nervous system. They are found in the brain, the spinal cord, in the nerves and ganglia of the peripheral nervous system. It is a specialized cell that conducts impulses through the nervous system and contains three major parts: cell body, dendrites, and an axon. It can have many dendrites but only one axon.
Bipolar cell	As a part of the retina, the bipolar cell exists between photoreceptors and ganglion cells. Bipolar cells are so-named as they have a central body from which two sets of processes arise. At one end, they form synapses with either a single cone cell, or a number of rod cells. At the other end, they form synapses with ganglion cells, which fire action potentials along the optic nerve (cranial nerve II). They effectively transfer information from rods and cones to ganglion cells.
Planum temporale	The planum temporale is the posterior superior surface of the superior temporal gyrus in the cerebrum. It is a highly lateralized brain structure involved with language.
Cochlear implant	A cochlear implant is a hearing device that can help people with certain kinds of hearing impairment or who have a severe to profound hearing loss. The implant works by using the tonotopic organization of the basilar membrane of the inner ear.
Neurologist	A physician who studies the nervous system, especially its structure, functions, and abnormalities is referred to as neurologist.
Paul Broca	Paul Broca became famous with his discovery of the speech center, the third circumvolution of the frontal lobe. He arrived at this discovery by studying the brains of aphasic patients (persons unable to talk).
Frontal lobe	The frontal lobe comprises four major folds of cortical tissue: the precentral gyrus, superior gyrus and the middle gyrus of the frontal gyri, the inferior frontal gyrus. It has been found to play a part in impulse control, judgement, language, memory, motor function, problem solving, sexual behavior, socialization and spontaneity.
Aphasia	Aphasia is a loss or impairment of the ability to produce or comprehend language, due to brain damage. It is usually a result of damage to the language centers of the brain.
Right hemisphere	The brain is divided into left and right cerebral hemispheres. The right hemisphere of the cortex controls the left side of the body.
Lesion	A lesion is a non-specific term referring to abnormal tissue in the body. It can be caused by any disease process including trauma (physical, chemical, electrical), infection, neoplasm, metabolic and autoimmune.
Penfield	Penfield treated patients with severe epilepsy by destroying nerve cells in the brain. Before operating, he stimulated the brain with electrical probes while the patients were conscious on the operating table, and observed their responses. It allowed him to create maps of sensory and motor functions.
Electroencep-alogram	Electroencephalography is the neurophysiologic measurement of the electrical activity of the brain by recording from electrodes placed on the scalp, or in the special cases on the cortex. The resulting traces are known as an electroencephalogram and represent so-called brainwaves.
Electrode	Any device used to electrically stimulate nerve tissue or to record its activity is an electrode.
Sensation	Sensation is the first stage in the chain of biochemical and neurologic events that begins with the impinging of a stimulus upon the receptor cells of a sensory organ, which then leads

	to perception, the mental state that is reflected in statements like "I see a uniformly blue wall."
Motor cortex	Motor cortex refers to the section of cortex that lies in the frontal lobe, just across the central fissure from the sensory cortex. Neural impulses in the motor cortex are linked to muscular responses throughout the body.
Somatosensory cortex	The primary somatosensory cortex is across the central sulcus and behind the primary motor cortex configured to generally correspond with the arrangement of nearby motor cells related to specific body parts. It is the main sensory receptive area for the sense of touch.
Consciousness	The awareness of the sensations, thoughts, and feelings being experienced at a given moment is called consciousness.
Seizure	A seizure is a temporary alteration in brain function expressed as a changed mental state, tonic or clonic movements and various other symptoms. They are due to temporary abnormal electrical activity of a group of brain cells.
Positron emission tomography	Positron Emission Tomography measures emissions from radioactively labeled chemicals that have been injected into the bloodstream. The greatest benefit is that different compounds can show blood flow and oxygen and glucose metabolism in the tissues of the working brain.
Photon	A photon is a quantum of the electromagnetic field, for instance light. In some respects a photon acts as a particle, for instance when registered by the light sensitive device in a camera. It also acts like a wave, as when passing through the optics in a camera.
Occipital lobe	The occipital lobe is the smallest of four true lobes in the human brain. Located in the rearmost portion of the skull, the occipital lobe is part of the forebrain structure. It is the visual processing center.
Cerebrovascular disease	Cerebrovascular disease is damage to the blood vessels in the brain, possibly resulting in a stroke.
Visual cortex	The visual cortex is the general term applied to both the primary visual cortex and the visual association area. Anatomically, the visual cortex occupies the entire occipital lobe, the inferior temporal lobe (IT), posterior parts of the parietal lobe, and a few small regions in the frontal lobe.
Fixation	Fixation in abnormal psychology is the state where an individual becomes obsessed with an attachment to another human, animal or inanimate object. Fixation in vision refers to maintaining the gaze in a constant direction. .
Discrimination	In Learning theory, discrimination refers the ability to distinguish between a conditioned stimulus and other stimuli. It can be brought about by extensive training or differential reinforcement. In social terms, it is the denial of privileges to a person or a group on the basis of prejudice.
Stimulus	A change in an environmental condition that elicits a response is a stimulus.
Stroke	A stroke occurs when the blood supply to a part of the brain is suddenly interrupted by occlusion, by hemorrhage, or other causes
Short-term memory	Short-term memory is that part of memory which stores a limited amount of information for a limited amount of time (roughly 30-45 seconds). The second key concept associated with a short-term memory is that it has a finite capacity.
Instinct	Instinct is the word used to describe inherent dispositions towards particular actions. They are generally an inherited pattern of responses or reactions to certain kinds of situations.
Epigenesis	The emergence of new structures and functions, during the course of development is referred to as epigenesis.

Cognition	The intellectual processes through which information is obtained, transformed, stored, retrieved, and otherwise used is cognition.
Babbling	Babbling is a stage in child language acquisition, during which an infant appears to be experimenting with making the sounds of language, but not yet producing any recognizable words.
Darwin	Darwin achieved lasting fame as originator of the theory of evolution through natural selection. His book Expression of Emotions in Man and Animals is generally considered the first text on comparative psychology.
Insight	Insight refers to a sudden awareness of the relationships among various elements that had previously appeared to be independent of one another.
Hormone	A hormone is a chemical messenger from one cell (or group of cells) to another. The best known are those produced by endocrine glands, but they are produced by nearly every organ system. The function of hormones is to serve as a signal to the target cells; the action of the hormone is determined by the pattern of secretion and the signal transduction of the receiving tissue.
Testosterone	Testosterone is a steroid hormone from the androgen group. It is the principal male sex hormone and the "original" anabolic steroid.
Larynx	The larynx, or voicebox, is an organ in the neck of mammals involved in protection of the trachea and sound production. The larynx houses the vocal cords, and is situated at the point where the upper tract splits into the trachea and the esophagus.
Ecology	Ecology refers to the branch of biology that deals with the relationships between living organisms and their environment.
Retina	The retina is a thin layer of cells at the back of the eyeball. It is the part of the eye which converts light into nervous signals. The retina contains photoreceptor cells which receive the light; the resulting neural signals then undergo complex processing by other neurons of the retina, and are transformed into action potentials in retinal ganglion cells whose axons form the optic nerve.
Fovea	The fovea, a part of the eye, is a spot located in the center of the macula. The fovea is responsible for sharp central vision, which is necessary in humans for reading, watching television or movies, driving, and any activity where visual detail is of primary importance.
Gazzaniga	Gazzaniga worked under the guidance of Roger Sperry, with primary responsibility for initiating human split-brain research. In his subsequent work he has made important advances in our understanding of functional lateralization in the brain and how the cerebral hemispheres communicate with one another.
Neuropsychology	Neuropsychology is a branch of psychology that aims to understand how the structure and function of the brain relates to specific psychological processes.
Theories	Theories are logically self-consistent models or frameworks describing the behavior of a certain natural or social phenomenon. They are broad explanations and predictions concerning phenomena of interest.
Somatosensory system	Somatosensory system consists of the various sensory receptors that trigger the experiences labelled as touch or pressure, temperature, pain, and the sensations of muscle movement and joint position including posture, movement, and facial expression.
Vestibular system	The vestibular system, or balance system, is the sensory system that provides the dominant input about our movement and orientation in space. Together with the cochlea, the auditory organ, it is situated in the vestibulum in the inner ear.

Basal ganglia	The basal ganglia are a group of nuclei in the brain associated with motor and learning functions.
Somatosensory	Somatosensory system consists of the various sensory receptors that trigger the experiences labelled as touch or pressure, temperature, pain, and the sensations of muscle movement and joint position including posture, movement, and facial expression.
Motor neuron	A motor neuron is an efferent neuron that originates in the spinal cord and synapses with muscle fibers to facilitate muscle contraction and with muscle spindles to modify proprioceptive sensitivity.
Spinal cord	The spinal cord is a part of the vertebrate nervous system that is enclosed in and protected by the vertebral column (it passes through the spinal canal). It consists of nerve cells. The spinal cord carries sensory signals and motor innervation to most of the skeletal muscles in the body.
Homunculus	Homunculus is a term used in a number of ways to describe systems that are thought of as being run by a "little man" inside. For instance, the homunculus continues to be considered as one of the major theories on the origin of consciousness, that there is a part in the brain whose purpose is to be "you".
Cerebellum	The cerebellum is located in the inferior posterior portion of the head (the hindbrain), directly dorsal to the brainstem and pons, inferior to the occipital lobe. The cerebellum is a region of the brain that plays an important role in the integration of sensory perception and fine motor output.
Forebrain	The forebrain is the highest level of the brain. Key structures in the forebrain are the limbic system, thalamus, basal ganglia, hypothalamus, and cerebral cortex.
Syndrome	The term syndrome is the association of several clinically recognizable features, signs, symptoms, phenomena or characteristics which often occur together, so that the presence of one feature indicates the presence of the others.
Ganglion	A ganglion is a tissue mass that contains the dendrites and cell bodies (or "somas") of nerve cells, in most case ones belonging to the peripheral nervous system. Within the central nervous system such a mass is often called a nucleus.
Autism	Autism is a neurodevelopmental disorder that manifests itself in markedly abnormal social interaction, communication ability, patterns of interests, and patterns of behavior.
Affect	A subjective feeling or emotional tone often accompanied by bodily expressions noticeable to others is called affect.
Senses	The senses are systems that consist of a sensory cell type that respond to a specific kind of physical energy, and that correspond to a defined region within the brain where the signals are received and interpreted.

Projection	Attributing one's own undesirable thoughts, impulses, traits, or behaviors to others is referred to as projection.
Nervous system	The body's electrochemical communication circuitry, made up of billions of neurons is a nervous system.
Spinal cord	The spinal cord is a part of the vertebrate nervous system that is enclosed in and protected by the vertebral column (it passes through the spinal canal). It consists of nerve cells. The spinal cord carries sensory signals and motor innervation to most of the skeletal muscles in the body.
Neocortex	The neocortex is part of the cerebral cortex which covers most of the surface of the cerebral hemispheres including the frontal, parietal, occipital, and temporal lobes. Often seen as the hallmark of human intelligence, the role of this structure in the brain appears to be involved in conscious thought, spatial reasoning, and sensory perception.
Sensory receptor	A sensory receptor is a structure that recognizes a stimulus in the environment of an organism. In response to stimuli the sensory receptor initiates sensory transduction by creating graded potentials or action potentials in the same cell or in an adjacent one.
Motor neuron	A motor neuron is an efferent neuron that originates in the spinal cord and synapses with muscle fibers to facilitate muscle contraction and with muscle spindles to modify proprioceptive sensitivity.
Neuron	The neuron is the primary cell of the nervous system. They are found in the brain, the spinal cord, in the nerves and ganglia of the peripheral nervous system. It is a specialized cell that conducts impulses through the nervous system and contains three major parts: cell body, dendrites, and an axon. It can have many dendrites but only one axon.
Brain	The brain controls and coordinates most movement, behavior and homeostatic body functions such as heartbeat, blood pressure, fluid balance and body temperature. Functions of the brain are responsible for cognition, emotion, memory, motor learning and other sorts of learning. The brain is primarily made up of two types of cells: glia and neurons.
Frontal lobe	The frontal lobe comprises four major folds of cortical tissue: the precentral gyrus, superior gyrus and the middle gyrus of the frontal gyri, the inferior frontal gyrus. It has been found to play a part in impulse control, judgement, language, memory, motor function, problem solving, sexual behavior, socialization and spontaneity.
Basal ganglia	The basal ganglia are a group of nuclei in the brain associated with motor and learning functions.
Cerebellum	The cerebellum is located in the inferior posterior portion of the head (the hindbrain), directly dorsal to the brainstem and pons, inferior to the occipital lobe. The cerebellum is a region of the brain that plays an important role in the integration of sensory perception and fine motor output.
Motor cortex	Motor cortex refers to the section of cortex that lies in the frontal lobe, just across the central fissure from the sensory cortex. Neural impulses in the motor cortex are linked to muscular responses throughout the body.
Neurologist	A physician who studies the nervous system, especially its structure, functions, and abnormalities is referred to as neurologist.
Forebrain	The forebrain is the highest level of the brain. Key structures in the forebrain are the limbic system, thalamus, basal ganglia, hypothalamus, and cerebral cortex.
Evolutionary theory	Evolutionary theory is concerned with heritable variability rather than behavioral variations. Natural selection requirements: (1) natural variability within a species must exist, (2) only some individual differences are heritable, and (3) natural selection only

takes place when there is an interaction between the inborn attributes of organisms and the environment in which they live.

Feedback — Feedback refers to information returned to a person about the effects a response has had.

Lashley — Lashley failed to find a single biological locus of memory suggesting to him that memories were not localized to one part of the brain, but were widely distributed throughout the cortex.

Shaping — The concept of reinforcing successive, increasingly accurate approximations to a target behavior is called shaping. The target behavior is broken down into a hierarchy of elemental steps, each step more sophisticated then the last. By successively reinforcing each of the the elemental steps, a form of differential reinforcement, until that step is learned while extinguishing the step below, the target behavior is gradually achieved.

Prefrontal cortex — The prefrontal cortex is the anterior part of the frontal lobes of the brain, lying in front of the motor and associative areas. It has been implicated in planning complex cognitive behaviors, personality expression and moderating correct social behavior. The prefrontal cortex continues to develop until around age 6.

Lesion — A lesion is a non-specific term referring to abnormal tissue in the body. It can be caused by any disease process including trauma (physical, chemical, electrical), infection, neoplasm, metabolic and autoimmune.

Cerebral cortex — The cerebral cortex is the outermost layer of the cerebrum and has a grey color. It is made up of four lobes and it is involved in many complex brain functions including memory, perceptual awareness, "thinking", language and consciousness. The cerebral cortex receives sensory information from many different sensory organs eg: eyes, ears, etc. and processes the information.

Somatosensory — Somatosensory system consists of the various sensory receptors that trigger the experiences labelled as touch or pressure, temperature, pain, and the sensations of muscle movement and joint position including posture, movement, and facial expression.

Electrode — Any device used to electrically stimulate nerve tissue or to record its activity is an electrode.

Adaptive behavior — An adaptive behavior increases the probability of the individual or organism to survive or exist within its environment.

Species-typical behavior — Species-typical behavior refers to any behavior pattern that is so characteristic of a given species of animal that it can be used to help identify that species.

Species — Species refers to a reproductively isolated breeding population.

Innate — Innate behavior is not learned or influenced by the environment, rather, it is present or predisposed at birth.

Stimulus — A change in an environmental condition that elicits a response is a stimulus.

Cerebral palsy — Cerebral palsy is a group of permanent disorders associated with developmental brain injuries that occur during fetal development, birth, or shortly after birth. It is characterized by a disruption of motor skills, with symptoms such as spasticity, paralysis, or seizures.

Brain trauma — Brain Trauma, also called acquired brain injury, intracranial injury, or simply head injury, occurs when a sudden trauma causes damage to the brain.

Autism — Autism is a neurodevelopmental disorder that manifests itself in markedly abnormal social interaction, communication ability, patterns of interests, and patterns of behavior.

Compensation — In personaility, compensation, according to Adler, is an effort to overcome imagined or real

inferiorities by developing one's abilities.

Sensation Sensation is the first stage in the chain of biochemical and neurologic events that begins with the impinging of a stimulus upon the receptor cells of a sensory organ, which then leads to perception, the mental state that is reflected in statements like "I see a uniformly blue wall."

Kanner Kanner was known for his work related to autism. He was the first physician in the United States to be identified as a child psychiatrist and his first textbook, Child Psychiatry in 1935, was the first English language textbook to focus on the psychiatric problems of children.

Attachment Attachment is the tendency to seek closeness to another person and feel secure when that person is present.

Nucleus In neuroanatomy, a cluster of cell bodies of neurons within the central nervous system is a nucleus.

Limbic system The limbic system is a group of brain structures that are involved in various emotions such as aggression, fear, pleasure and also in the formation of memory. The limbic system affects the endocrine system and the autonomic nervous system. It consists of several subcortical structures located around the thalamus.

Control subjects Control subjects are participants in an experiment who do not receive the treatment effect but for whom all other conditions are held comparable to those of experimental subjects.

Trait An enduring personality characteristic that tends to lead to certain behaviors is called a trait. The term trait also means a genetically inherited feature of an organism.

Fraternal twins Fraternal twins usually occur when two fertilized eggs are implanted in the uterine wall at the same time. The two eggs form two zygotes, and these twins are therefore also known as dizygotic. Dizygotic twins are no more similar genetically than any siblings.

Rubella An infectious disease that, if contracted by the mother during the first three months of pregnancy, has a high risk of causing mental retardation and physical deformity in the child is called rubella.

Pons The pons is a knob on the brain stem. It is part of the autonomic nervous system, and relays sensory information between the cerebellum and cerebrum. Some theories posit that it has a role in dreaming.

Measles Measles, also known as rubeola, is a common disease caused by a virus of the genus Morbillivirus. Complications with measles are relatively common, ranging from relatively common and less serious diarrhea, to pneumonia and encephalitis. Complications are usually more severe amongst infants and adults who catch the virus.

Gene A gene is an ultramicroscopic area of the chromosome. It is the smallest physical unit of the DNA molecule that carries a piece of hereditary information.

Receptor A sensory receptor is a structure that recognizes a stimulus in the internal or external environment of an organism. In response to stimuli the sensory receptor initiates sensory transduction by creating graded potentials or action potentials in the same cell or in an adjacent one.

Tactile Pertaining to the sense of touch is referred to as tactile.

Reflex A simple, involuntary response to a stimulus is referred to as reflex. Reflex actions originate at the spinal cord rather than the brain.

Nerve A nerve is an enclosed, cable-like bundle of nerve fibers or axons, which includes the glia that ensheath the axons in myelin. Neurons are sometimes called nerve cells, though this term

	is technically imprecise since many neurons do not form nerves.
Schwann cell	A Schwann cell is a variety of neuroglia that wraps around axons in the peripheral nervous system, forming the myelin sheath. The nervous system depends crucially on this sheath for insulation and an increase in impulse speed.
Myelin	Myelin is an electrically insulating fatty layer that surrounds the axons of many neurons, especially those in the peripheral nervous system. The main consequence of a myelin sheath is an increase in the speed at which impulses propagate along the myelinated fiber. The sheath continues to develop throughout childhood.
Axon	An axon, or "nerve fiber," is a long slender projection of a nerve cell, or "neuron," which conducts electrical impulses away from the neuron's cell body or soma.
Penfield	Penfield treated patients with severe epilepsy by destroying nerve cells in the brain. Before operating, he stimulated the brain with electrical probes while the patients were conscious on the operating table, and observed their responses. It allowed him to create maps of sensory and motor functions.
Gyrus	A gyrus is a ridge on the cerebral cortex. It is generally surrounded by one or more sulci.
Homunculus	Homunculus is a term used in a number of ways to describe systems that are thought of as being run by a "little man" inside. For instance, the homunculus continues to be considered as one of the major theories on the origin of consciousness, that there is a part in the brain whose purpose is to be "you".
Relative size	Relative size is a monocular cue where an an automobile that is close to us seems larger than one that is far away. The visual system exploits the relative size of similar (or familiar) objects to judge distance.
Right hemisphere	The brain is divided into left and right cerebral hemispheres. The right hemisphere of the cortex controls the left side of the body.
Left hemisphere	The left hemisphere of the cortex controls the right side of the body, coordinates complex movements, and, in 95% of people, controls the production of speech and written language.
Interneuron	An interneuron (also called relay neuron or association neuron) is a neuron that communicates only to other neurons.
Synapse	A synapse is specialized junction through which cells of the nervous system signal to one another and to non-neuronal cells such as muscles or glands.
Autonomic nervous system	A division of the peripheral nervous system, the autonomic nervous system, regulates glands and activities such as heartbeat, respiration, digestion, and dilation of the pupils. It is responsible for homeostasis, maintaining a relatively constant internal environment .
Neurotransmitter	A neurotransmitter is a chemical that is used to relay, amplify and modulate electrical signals between a neurons and another cell.
Acetylcholine	The chemical compound acetylcholine was the first neurotransmitter to be identified. It plays a role in learning, memory, and rapid eye movement sleep and causes the skeletal muscle fibers to contract.
Neuroscience	A field that combines the work of psychologists, biologists, biochemists, medical researchers, and others in the study of the structure and function of the nervous system is neuroscience.
Coding	In senation, coding is the process by which information about the quality and quantity of a stimulus is preserved in the pattern of action potentials sent through sensory neurons to the central nervous system.

Displacement	An unconscious defense mechanism in which the individual directs aggressive or sexual feelings away from the primary object to someone or something safe is referred to as displacement. Displacement in linguistics is simply the ability to talk about things not present.
Population	Population refers to all members of a well-defined group of organisms, events, or things.
Hypothesis	A specific statement about behavior or mental processes that is testable through research is a hypothesis.
Mutation	Mutation is a permanent, sometimes transmissible (if the change is to a germ cell) change to the genetic material (usually DNA or RNA) of a cell. They can be caused by copying errors in the genetic material during cell division and by exposure to radiation, chemicals, or viruses, or can occur deliberately under cellular control during the processes such as meiosis or hypermutation.
Affect	A subjective feeling or emotional tone often accompanied by bodily expressions noticeable to others is called affect.
Anatomy	Anatomy is the branch of biology that deals with the structure and organization of living things. It can be divided into animal anatomy (zootomy) and plant anatomy (phytonomy). Major branches of anatomy include comparative anatomy, histology, and human anatomy.
Midbrain	Located between the hindbrain and forebrain, a region in which many nerve-fiber systems ascend and descend to connect the higher and lower portions of the brain is referred to as midbrain. It is archipallian in origin, meaning its general architecture is shared with the most ancient of vertebrates. Dopamine produced in the subtantia nigra plays a role in motivation and habituation of species from humans to the most elementary animals such as insects.
Temporal lobe	The temporal lobe is part of the cerebrum. It lies at the side of the brain, beneath the lateral or Sylvian fissure. Adjacent areas in the superior, posterior and lateral parts of the temporal lobe are involved in high-level auditory processing.
Amygdala	Located in the brain's medial temporal lobe, the almond-shaped amygdala is believed to play a key role in the emotions. It forms part of the limbic system and is linked to both fear responses and pleasure. Its size is positively correlated with aggressive behavior across species.
Substantia nigra	The substantia nigra is a portion of the midbrain thought to be involved in certain aspects of movement and attention. Degeneration of cells in this region is the principle pathology that underlies Parkinson's disease.
Caudate nucleus	The caudate nucleus is a telencephalic nucleus, one of the input nuclei of the basal ganglia; involved with control of voluntary movement in the brain.
Thalamus	An area near the center of the brain involved in the relay of sensory information to the cortex and in the functions of sleep and attention is the thalamus.
Syndrome	The term syndrome is the association of several clinically recognizable features, signs, symptoms, phenomena or characteristics which often occur together, so that the presence of one feature indicates the presence of the others.
Tics	Tics are a repeated, impulsive action, almost reflexive in nature, which the person feels powerless to control or avoid.
Haloperidol	Haloperidol is a conventional butyrophenone antipsychotic drug. It posesses a strong activity against delusions and hallucinations, most likely due to an effective dopaminergic receptor blockage in the mesocortex and the limbic system of the brain.

Dopamine	Dopamine is critical to the way the brain controls our movements and is a crucial part of the basal ganglia motor loop. It is commonly associated with the 'pleasure system' of the brain, providing feelings of enjoyment and reinforcement to motivate us to do, or continue doing, certain activities.
Society	The social sciences use the term society to mean a group of people that form a semi-closed (or semi-open) social system, in which most interactions are with other individuals belonging to the group.
Attention	Attention is the cognitive process of selectively concentrating on one thing while ignoring other things. Psychologists have labeled three types of attention: sustained attention, selective attention, and divided attention.
Somatosensory system	Somatosensory system consists of the various sensory receptors that trigger the experiences labelled as touch or pressure, temperature, pain, and the sensations of muscle movement and joint position including posture, movement, and facial expression.
Somatosensation	Somatosensation consists of the various sensory receptors that trigger the experiences labelled as touch or pressure, temperature (warm or cold), pain (including itch and tickle), and the sensations of muscle movement and joint position including posture, movement, and facial expression (collectively also called proprioception).
Senses	The senses are systems that consist of a sensory cell type that respond to a specific kind of physical energy, and that correspond to a defined region within the brain where the signals are received and interpreted.
Vestibular system	The vestibular system, or balance system, is the sensory system that provides the dominant input about our movement and orientation in space. Together with the cochlea, the auditory organ, it is situated in the vestibulum in the inner ear.
Middle ear	The middle ear consists of the eardrum, hammer, anvil, and stirrup.
Sensory neuron	A sensory neuron is an afferent nerve cell within the nervous system responsible for converting external stimuli from the organism's environment into internal electrical impulses. It carries messages from a sensory organ, through a nerve, into the brain or spinal cord.
Dendrite	A dendrite is a slender, typically branched projection of a nerve cell, or "neuron," which conducts the electrical stimulation received from other cells to the body or soma of the cell from which it projects. This stimulation arrives through synapses, which typically are located near the tips of the dendrites and away from the soma.
Perception	Perception is the process of acquiring, interpreting, selecting, and organizing sensory information.
Adaptation	Adaptation is a lowering of sensitivity to a stimulus following prolonged exposure to that stimulus. Behavioral adaptations are special ways a particular organism behaves to survive in its natural habitat.
Action potential	The electrical impulse that provides the basis for the conduction of a neural impulse along an axon of a neuron is the action potential. When a biological cell or patch of membrane undergoes an action potential, or electrical excitation, the polarity of the transmembrane voltage swings rapidly from negative to positive and back.
Proprioception	Proprioception is the sense of the position of parts of the body, relative to other neighboring parts of the body.
Peptides	Peptides are the family of molecules formed from the linking, in a defined order, of various amino acids.

Golgi	Golgi discovered a method of staining nervous tissue which would stain a limited number of cells at random, in their entirety. This enabled him to view the paths of nerve cells in the brain for the first time. He called his discovery the black reaction. It is now known universally as the Golgi stain.
Ion channel	An Ion channel is a pore-forming protein that help establish the small voltage gradient that exists across the membrane of all living cells, by controlling the flow of ions. They are present in the membranes that surround all biological cells.
Ganglion	A ganglion is a tissue mass that contains the dendrites and cell bodies (or "somas") of nerve cells, in most case ones belonging to the peripheral nervous system. Within the central nervous system such a mass is often called a nucleus.
Somatosensory cortex	The primary somatosensory cortex is across the central sulcus and behind the primary motor cortex configured to generally correspond with the arrangement of nearby motor cells related to specific body parts. It is the main sensory receptive area for the sense of touch.
Myelination	The process in which the nerve cells are covered and insulated with a layer of fat cells, which increases the speed at which information travels through the nervous system is referred to as myelination.
Sensory nerves	Sensory nerves bring impulses toward the central nervous system.
Central nervous system	The vertebrate central nervous system consists of the brain and spinal cord.
Morphine	Morphine, the principal active agent in opium, is a powerful opioid analgesic drug. According to recent research, it may also be produced naturally by the human brain. Morphine is usually highly addictive, and tolerance and physical and psychological dependence develop quickly.
Opioid	An opioid is any agent that binds to opioid receptors, found principally in the central nervous system and gastrointestinal tract.
Semicircular canals	The semicircular canals are three half-circular, interconnected tubes located inside each ear that are the equivalent of three gyroscopes located in three planes perpendicular (at right angles) to each other.
Otolith	The otolith organs (the utricle and the saccule) are structures in the inner ear that are sensitive to gravity and linear acceleration. Because of their orientation in the head, the utricle is sensitive to a change in horizontal movement, and the saccule gives information about vertical acceleration (such as when in an elevator).
Endolymph	Endolymph is the fluid contained in the membranous labyrinth of the inner ear. Disruption of the endolymph due to jerky movements (like driving over bumps while riding in a car) can cause motion sickness.
Hair cells	Hair cells are the sensory cells of both the auditory system and the vestibular system. The auditory hair cells are located within the organ of Corti on a thin basilar membrane in the cochlea of the inner ear.
Ganglion cell	A ganglion cell is a type of neuron located in the retina of the eye that receives visual information from photoreceptors via various intermediate cells such as bipolar cells, amacrine cells, and horizontal cells. Retinal ganglion cells' axons are myelinated.
Parietal lobe	The parietal lobe is positioned above (superior to) the occipital lobe and behind (posterior to) the frontal lobe. It plays important roles in integrating sensory information from various senses, and in the manipulation of objects.
Construct	A generalized concept, such as anxiety or gravity, is a construct.
Visual cortex	The visual cortex is the general term applied to both the primary visual cortex and the

	visual association area. Anatomically, the visual cortex occupies the entire occipital lobe, the inferior temporal lobe (IT), posterior parts of the parietal lobe, and a few small regions in the frontal lobe.
Skin receptors	Skin receptors refer to sensory organs for touch, pressure, pain, cold, and warmth.
Discrimination	In Learning theory, discrimination refers the ability to distinguish between a conditioned stimulus and other stimuli. It can be brought about by extensive training or differential reinforcement. In social terms, it is the denial of privileges to a person or a group on the basis of prejudice.
Threshold	In general, a threshold is a fixed location or value where an abrupt change is observed. In the sensory modalities, it is the minimum amount of stimulus energy necessary to elicit a sensory response.
Animal model	An animal model usually refers to a non-human animal with a disease that is similar to a human condition.
Insight	Insight refers to a sudden awareness of the relationships among various elements that had previously appeared to be independent of one another.
Ventral stream	The primate visual system consists of about thirty areas of the cerebral cortex called the visual cortex. The visual cortex is divided into the ventral stream and the dorsal stream. The ventral stream is associated with object recognition and form representation.
Species-specific behavior	An adaptive, innate response pattern typical of a species in a given situation is a species-specific behavior. Whether the behavior is modifiable through experience is debated.
Dorsal stream	The dorsal stream is a pathway for visual information which flows through the visual cortex, the part of the brain which provides visual processing. It is involved in spatial awareness: recognizing where objects are in space.
Disuse	Disuse refers to theory that memory traces weaken when memories are not periodically used or retrieved.
Theories	Theories are logically self-consistent models or frameworks describing the behavior of a certain natural or social phenomenon. They are broad explanations and predictions concerning phenomena of interest.

Hypothalamus	The hypothalamus is a region of the brain located below the thalamus, forming the major portion of the ventral region of the diencephalon and functioning to regulate certain metabolic processes and other autonomic activities.
Tumor	A tumor is an abnormal growth that when located in the brain can either be malignant and directly destroy brain tissue, or be benign and disrupt functioning by increasing intracranial pressure.
Brain	The brain controls and coordinates most movement, behavior and homeostatic body functions such as heartbeat, blood pressure, fluid balance and body temperature. Functions of the brain are responsible for cognition, emotion, memory, motor learning and other sorts of learning. The brain is primarily made up of two types of cells: glia and neurons.
Suicide	Suicide behavior is rare in childhood but escalates in adolescence. The suicide rate increases in a linear fashion from adolescence through late adulthood.
Sensory deprivation	Sensory deprivation is the deliberate reduction or removal of stimuli from one or more of the senses. Though short periods of sensory deprivation can be relaxing, extended deprivation can result in extreme anxiety, hallucinations, bizarre thoughts, depression, and antisocial behavior.
Pathology	Pathology is the study of the processes underlying disease and other forms of illness, harmful abnormality, or dysfunction.
Free will	The idea that human beings are capable of freely making choices or decisions is free will.
Hebb	Hebb demonstrated that the rearing of rats in an enriched environment could alter neural development and that sensory - neural connections were shaped by experience. He is famous for developing the concept of neural nets. He also believed that learning early in life is of the incremental variety, whereas later it is cognitive, insightful, and more all-or-none.
Incentive	An incentive is what is expected once a behavior is performed. An incentive acts as a reinforcer.
Physiological psychology	Physiological psychology refers to the study of the physiological mechanisms, in the brain and elsewhere, that mediate behavior and psychological experiences.
Harlow	Harlow and his famous wire and cloth surrogate mother monkey studies demonstrated that the need for affection created a stronger bond between mother and infant than did physical needs. He also found that the more discrimination problems the monkeys solved, the better they became at solving them.
Attention	Attention is the cognitive process of selectively concentrating on one thing while ignoring other things. Psychologists have labeled three types of attention: sustained attention, selective attention, and divided attention.
Theories	Theories are logically self-consistent models or frameworks describing the behavior of a certain natural or social phenomenon. They are broad explanations and predictions concerning phenomena of interest.
Drive Theory	Drive Theory states that due to the unpredictable nature of people, a person performing a task rarely knows for certain what others are going to do in response. This creates a state of arousal that in some cases will facilitate performance, while in other cases it will inhibit performance.
Analogy	An analogy is a comparison between two different things, in order to highlight some form of similarity. Analogy is the cognitive process of transferring information from a particular subject to another particular subject.
Stimulus	A change in an environmental condition that elicits a response is a stimulus.

Hormone	A hormone is a chemical messenger from one cell (or group of cells) to another. The best known are those produced by endocrine glands, but they are produced by nearly every organ system. The function of hormones is to serve as a signal to the target cells; the action of the hormone is determined by the pattern of secretion and the signal transduction of the receiving tissue.
Correlation	A statistical technique for determining the degree of association between two or more variables is referred to as correlation.
Neuron	The neuron is the primary cell of the nervous system. They are found in the brain, the spinal cord, in the nerves and ganglia of the peripheral nervous system. It is a specialized cell that conducts impulses through the nervous system and contains three major parts: cell body, dendrites, and an axon. It can have many dendrites but only one axon.
Plasticity	The capacity for modification and change is referred to as plasticity.
Innate	Innate behavior is not learned or influenced by the environment, rather, it is present or predisposed at birth.
Innate releasing mechanism	An innate releasing mechanism is a hypothetical innate system within an animal that responds to a stimulus in the environment to produce a pre-programmed stereotyped behavior or fixed action pattern.
Nervous system	The body's electrochemical communication circuitry, made up of billions of neurons is a nervous system.
Species	Species refers to a reproductively isolated breeding population.
Discrimination	In Learning theory, discrimination refers the ability to distinguish between a conditioned stimulus and other stimuli. It can be brought about by extensive training or differential reinforcement. In social terms, it is the denial of privileges to a person or a group on the basis of prejudice.
Natural selection	Natural selection is a process by which biological populations are altered over time, as a result of the propagation of heritable traits that affect the capacity of individual organisms to survive and reproduce.
Brain circuits	Neurotransmitter currents or neural pathways in the brain are referred to as brain circuits.
Trait	An enduring personality characteristic that tends to lead to certain behaviors is called a trait. The term trait also means a genetically inherited feature of an organism.
Gene	A gene is an ultramicroscopic area of the chromosome. It is the smallest physical unit of the DNA molecule that carries a piece of hereditary information.
Evolutionary psychology	Evolutionary psychology proposes that cognition and behavior can be better understood in light of evolutionary history.
Insight	Insight refers to a sudden awareness of the relationships among various elements that had previously appeared to be independent of one another.
Self-esteem	Self-esteem refers to a person's subjective appraisal of himself or herself as intrinsically positive or negative to some degree.
Evolution	Commonly used to refer to gradual change, evolution is the change in the frequency of alleles within a population from one generation to the next. This change may be caused by different mechanisms, including natural selection, genetic drift, or changes in population (gene flow).
Population	Population refers to all members of a well-defined group of organisms, events, or things.
Punishment	Punishment is the addtion of a stimulus that reduces the frequency of a response, or the removal of a stimulus that results in a reduction of the response.

Society	The social sciences use the term society to mean a group of people that form a semi-closed (or semi-open) social system, in which most interactions are with other individuals belonging to the group.
Evolutionary theory	Evolutionary theory is concerned with heritable variability rather than behavioral variations. Natural selection requirements: (1) natural variability within a species must exist, (2) only some individual differences are heritable, and (3) natural selection only takes place when there is an interaction between the inborn attributes of organisms and the environment in which they live.
Learning	Learning is a relatively permanent change in behavior that results from experience. Thus, to attribute a behavioral change to learning, the change must be relatively permanent and must result from experience.
Skinner	Skinner conducted research on shaping behavior through positive and negative reinforcement, and demonstrated operant conditioning, a technique which he developed in contrast with classical conditioning.
Reinforcer	In operant conditioning, a reinforcer is any stimulus that increases the probability that a preceding behavior will occur again. In Classical Conditioning, the unconditioned stimulus (US) is the reinforcer.
Reinforcement	In operant conditioning, reinforcement is any change in an environment that (a) occurs after the behavior, (b) seems to make that behavior re-occur more often in the future and (c) that reoccurence of behavior must be the result of the change.
Anxiety	Anxiety is a complex combination of the feeling of fear, apprehension and worry often accompanied by physical sensations such as palpitations, chest pain and/or shortness of breath.
Illusion	An illusion is a distortion of a sensory perception.
Skinner box	An operant conditioning chamber, or Skinner box, is an experimental apparatus used to study conditioning in animals. Chambers have at least one operandum that can automatically detect the occurrence of a behavioral response or action. The other minimal requirement of a conditioning chamber is that it have a means of delivering a primary reinforcer or unconditioned stimulus like food or water.
Preparedness	The species-specific biological predisposition to learn in certain ways is called preparedness.
Receptor	A sensory receptor is a structure that recognizes a stimulus in the internal or external environment of an organism. In response to stimuli the sensory receptor initiates sensory transduction by creating graded potentials or action potentials in the same cell or in an adjacent one.
Schematic representation	The representation of objects in terms of real or potential interactions with other objects is called a schematic representation.
Nerve	A nerve is an enclosed, cable-like bundle of nerve fibers or axons, which includes the glia that ensheath the axons in myelin. Neurons are sometimes called nerve cells, though this term is technically imprecise since many neurons do not form nerves.
Olfactory system	The olfactory system is the sensory system used for the sense of smell. The olfactory system is often spoken of along with the gustatory system as the chemosensory senses because both transduce chemical signals into perception
Pituitary gland	The pituitary gland is an endocrine gland about the size of a pea that sits in the small, bony cavity at the base of the brain. The pituitary gland secretes hormones regulating a wide variety of bodily activities, including trophic hormones that stimulate other endocrine

	glands.
Limbic system	The limbic system is a group of brain structures that are involved in various emotions such as aggression, fear, pleasure and also in the formation of memory. The limbic system affects the endocrine system and the autonomic nervous system. It consists of several subcortical structures located around the thalamus.
Frontal lobe	The frontal lobe comprises four major folds of cortical tissue: the precentral gyrus, superior gyrus and the middle gyrus of the frontal gyri, the inferior frontal gyrus. It has been found to play a part in impulse control, judgement, language, memory, motor function, problem solving, sexual behavior, socialization and spontaneity.
Anatomy	Anatomy is the branch of biology that deals with the structure and organization of living things. It can be divided into animal anatomy (zootomy) and plant anatomy (phytonomy). Major branches of anatomy include comparative anatomy, histology, and human anatomy.
Axon	An axon, or "nerve fiber," is a long slender projection of a nerve cell, or "neuron," which conducts electrical impulses away from the neuron's cell body or soma.
Autonomic nervous system	A division of the peripheral nervous system, the autonomic nervous system, regulates glands and activities such as heartbeat, respiration, digestion, and dilation of the pupils. It is responsible for homeostasis, maintaining a relatively constant internal environment .
Optic chiasm	Optic chiasm refers to the point at which the optic nerves from the inside half of each eye cross over and then project to the opposite half of the brain.
Optic tract	The optic tract is a part of the visual system in the brain. It is a continuation of the optic nerve and runs from the optic chiasm (where half of the information from each eye crosses sides, and half stays on the same side) to the lateral geniculate nucleus.
Optic nerve	The optic nerve is the nerve that transmits visual information from the retina to the brain. The optic nerve is composed of retinal ganglion cell axons and support cells.
Forebrain	The forebrain is the highest level of the brain. Key structures in the forebrain are the limbic system, thalamus, basal ganglia, hypothalamus, and cerebral cortex.
Nucleus	In neuroanatomy, a cluster of cell bodies of neurons within the central nervous system is a nucleus.
Basal ganglia	The basal ganglia are a group of nuclei in the brain associated with motor and learning functions.
Projection	Attributing one's own undesirable thoughts, impulses, traits, or behaviors to others is referred to as projection.
Peptides	Peptides are the family of molecules formed from the linking, in a defined order, of various amino acids.
Dopamine	Dopamine is critical to the way the brain controls our movements and is a crucial part of the basal ganglia motor loop. It is commonly associated with the 'pleasure system' of the brain, providing feelings of enjoyment and reinforcement to motivate us to do, or continue doing, certain activities.
Oxytocin	Oxytocin is synthesized in magnocellular neurosecretory cells in the hypothalamus and released by the posterior lobe of the pituitary gland. It is involved in the facilitation of birth and breastfeeding as well as in bonding.
Action potential	The electrical impulse that provides the basis for the conduction of a neural impulse along an axon of a neuron is the action potential. When a biological cell or patch of membrane undergoes an action potential, or electrical excitation, the polarity of the transmembrane voltage swings rapidly from negative to positive and back.

Growth hormone	Growth hormone is a polypeptide hormone synthesised and secreted by the anterior pituitary gland which stimulates growth and cell reproduction in humans and other vertebrate animals.
Adrenal cortex	Adrenal cortex refers to the outer layer of the adrenal glands, which produces hormones that affect salt intake, reactions to stress, and sexual development.
Thyroid Gland	The thyroid gland is an endocrine gland. It is situated on the front side of the neck just below the Adam's apple. The primary function of the thyroid is production of hormones such as thyroxine.
Gonads	The gonads are the organs that make gametes. Gametes are haploid germ cells. For example, sperm and egg cells are gametes. In the male the gonads are the testicles, and in the female the gonads are the ovaries.
Gland	A gland is an organ in an animal's body that synthesizes a substance for release such as hormones, often into the bloodstream or into cavities inside the body or its outer surface.
Thyroid	In anatomy, the thyroid is the largest endocrine gland in the body. The primary function of the thyroid is production of hormones.
Axon terminal	A swelling at the end of an axon that is designed to release a chemical substance onto another neuron, muscle cell, or gland cell is called the axon terminal.
Tactile	Pertaining to the sense of touch is referred to as tactile.
Paul Broca	Paul Broca became famous with his discovery of the speech center, the third circumvolution of the frontal lobe. He arrived at this discovery by studying the brains of aphasic patients (persons unable to talk).
Emotion	An emotion is a mental states that arise spontaneously, rather than through conscious effort. They are often accompanied by physiological changes.
Amygdala	Located in the brain's medial temporal lobe, the almond-shaped amygdala is believed to play a key role in the emotions. It forms part of the limbic system and is linked to both fear responses and pleasure. Its size is positively correlated with aggressive behavior across species.
Prefrontal cortex	The prefrontal cortex is the anterior part of the frontal lobes of the brain, lying in front of the motor and associative areas. It has been implicated in planning complex cognitive behaviors, personality expression and moderating correct social behavior. The prefrontal cortex continues to develop until around age 6.
Hippocampus	The hippocampus is a part of the brain located inside the temporal lobe. It forms a part of the limbic system and plays a part in memory and navigation.
Gyrus	A gyrus is a ridge on the cerebral cortex. It is generally surrounded by one or more sulci.
Cingulate cortex	The part of the limbic system that is believed to process cognitive information in emotion is the cingulate cortex. The cingulate cortex is part of the brain and situated roughly in the middle of the cortex.
Thalamus	An area near the center of the brain involved in the relay of sensory information to the cortex and in the functions of sleep and attention is the thalamus.
Species-typical behavior	Species-typical behavior refers to any behavior pattern that is so characteristic of a given species of animal that it can be used to help identify that species.
Lesion	A lesion is a non-specific term referring to abnormal tissue in the body. It can be caused by any disease process including trauma (physical, chemical, electrical), infection, neoplasm, metabolic and autoimmune.
Motor cortex	Motor cortex refers to the section of cortex that lies in the frontal lobe, just across the

	central fissure from the sensory cortex. Neural impulses in the motor cortex are linked to muscular responses throughout the body.
Mammillary body	The mammillary body is a pair of small round bodies in the brain forming part of the limbic system. Symptoms from damage to the mammillary bodies can include impaired memory, also called anterograde amnesia; this suggests that the mammillary bodies might be important for memory.
Central nervous system	The vertebrate central nervous system consists of the brain and spinal cord.
Schizophrenia	Schizophrenia is characterized by persistent defects in the perception or expression of reality. A person suffering from untreated schizophrenia typically demonstrates grossly disorganized thinking, and may also experience delusions or auditory hallucinations
Motivation	In psychology, motivation is the driving force (desire) behind all actions of an organism.
Executive function	The processes involved in regulating attention and in determining what to do with information just gathered or retrieved from long-term memory, is referred to as the executive function.
Eating disorders	Psychological disorders characterized by distortion of the body image and gross disturbances in eating patterns are called eating disorders.
Amino acid	Amino acid is the basic structural building unit of proteins. They form short polymer chains called peptides or polypeptides which in turn form structures called proteins.
Pancreas	The pancreas is a retroperitoneal organ that serves two functions: it produces juice containing digestive enzymes; and it produces several important hormones including insulin, glucagon, and several other hormones.
Gall	Gall most noted for introducing phrenology, was correct in assigning the brain the role of the seat of mental activities, and although he was wrong in detail due to a faulty methodology, the possibility of the localization of brain functions is widely accepted today.
Glucose	Glucose, a simple monosaccharide sugar, is one of the most important carbohydrates and is used as a source of energy in animals and plants. Glucose is one of the main products of photosynthesis and starts respiration.
Protein	A protein is a complex, high-molecular-weight organic compound that consists of amino acids joined by peptide bonds. It is essential to the structure and function of all living cells and viruses. Many are enzymes or subunits of enzymes.
Liver	The liver plays a major role in metabolism and has a number of functions in the body including detoxification, glycogen storage and plasma protein synthesis. It also produces bile, which is important for digestion. The liver converts most carbohydrates, proteing, and fats into glucose.
Gastrointestinal tract	The gastrointestinal tract is the system of organs within multicellular animals which takes in food, digests it to extract energy and nutrients, and expels the remaining waste.
Satiety	Satiety refers to the state of being satisfied; fullness.
Steroid	A steroid is a lipid characterized by a carbon skeleton with four fused rings. Different steroids vary in the functional groups attached to these rings. Hundreds of distinct steroids have been identified in plants and animals. Their most important role in most living systems is as hormones.
Ventromedial hypothalamus	Ventromedial hypothalamus acts as a satiety center and, when activated, signals an animal to stop eating; when destroyed, the animal overeats, becoming obese.
Obesity	The state of being more than 20 percent above the average weight for a person of one's height

	is called obesity.
Androgen insensitivity	Androgen insensitivity syndrome is a set of disorders of sexual differentiation that results from mutations of the gene encoding the androgen receptor. If androgen insensitivity is total, XY males develop as females in the sense that their bodies look completely female and they develop a female gender identity.
Androgenital syndrome	A disorder in which genetic females become masculinized as a result of prenatal exposure to male hormones is androgenital syndrome.
Testes	Testes are the male reproductive glands or gonads; this is where sperm develop and are stored.
Fetus	A fetus develops from the end of the 8th week of pregnancy (when the major structures have formed), until birth.
Syndrome	The term syndrome is the association of several clinically recognizable features, signs, symptoms, phenomena or characteristics which often occur together, so that the presence of one feature indicates the presence of the others.
Estrogen	Estrogen is a group of steroid compounds that function as the primary female sex hormone. They are produced primarily by developing follicles in the ovaries, the corpus luteum and the placenta.
Y chromosome	The Y chromosome is one of the two sex chromosomes in humans and most other mammals. The sex chromosomes are one of the 23 pairs of human chromosomes. The Y chromosome contains the fewest genes of any of the chromosomes. It contains the genes that cause testis development, thus determining maleness. It is usually contributed by the father.
Congenital adrenal hyperplasia	Congenital adrenal hyperplasia refers to a group of inherited disorders relating to the adrenal glands, characterized by a deficiency in the hormones cortisol and aldosterone and an overproduction of androgen. Without these hormones, steroids are 'diverted' to becoming androgens, a form of male sex hormones. This causes early (or inappropriate) appearance of male characteristics.
Clitoris	Clitoris refers to an external female sex organ that is highly sensitive to sexual stimulation.
Penis	The penis is the external male copulatory organ and the external male organ of urination. In humans, the penis is homologous to the female clitoris, as it develops from the same embryonic structure. It is capable of erection for use in copulation.
Early childhood	Early childhood refers to the developmental period extending from the end of infancy to about 5 or 6 years of age; sometimes called the preschool years.
Phenotype	The phenotype of an individual organism is either its total physical appearance and constitution, or a specific manifestation of a trait, such as size or eye color, that varies between individuals. Phenotype is determined to some extent by genotype, or by the identity of the alleles that an individual carries at one or more positions on the chromosomes.
Androgen	Androgen is the generic term for any natural or synthetic compound, usually a steroid hormone, that stimulates or controls the development and maintenance of masculine characteristics in vertebrates by binding to androgen receptors.
Testosterone	Testosterone is a steroid hormone from the androgen group. It is the principal male sex hormone and the "original" anabolic steroid.
Psychiatrist	A psychiatrist is a physician who specializes in the diagnosis and treatment of psychological disorders.
Adolescence	The period of life bounded by puberty and the assumption of adult responsibilities is

	adolescence.
Masturbation	Masturbation is the manual excitation of the sexual organs, most often to the point of orgasm. It can refer to excitation either by oneself or by another, but commonly refers to such activities performed alone.
Physiological changes	Alterations in heart rate, blood pressure, perspiration, and other involuntary responses are physiological changes.
Depression	In everyday language depression refers to any downturn in mood, which may be relatively transitory and perhaps due to something trivial. This is differentiated from Clinical depression which is marked by symptoms that last two weeks or more and are so severe that they interfere with daily living.
James-Lange	The James-Lange theory concerns the origin and nature of emotions as developed independently by two 19th-century scholars. As a response to experiences in the world, the autonomic nervous system creates physiological events. Emotions are feelings which come about as a result of these physiological changes, rather than being their cause.
Spinal cord	The spinal cord is a part of the vertebrate nervous system that is enclosed in and protected by the vertebral column (it passes through the spinal canal). It consists of nerve cells. The spinal cord carries sensory signals and motor innervation to most of the skeletal muscles in the body.
Neuroscience	A field that combines the work of psychologists, biologists, biochemists, medical researchers, and others in the study of the structure and function of the nervous system is neuroscience.
Anxiety disorder	Anxiety disorder is a blanket term covering several different forms of abnormal anxiety, fear, phobia and nervous condition, that come on suddenly and prevent pursuing normal daily routines.
Phobia	A persistent, irrational fear of an object, situation, or activity that the person feels compelled to avoid is referred to as a phobia.
Affect	A subjective feeling or emotional tone often accompanied by bodily expressions noticeable to others is called affect.
Panic disorder	A panic attack is a period of intense fear or discomfort, typically with an abrupt onset and usually lasting no more than thirty minutes. The disorder is strikingly different from other types of anxiety, in that panic attacks are very sudden, appear to be unprovoked, and are often disabling. People who have repeated attacks, or feel severe anxiety about having another attack are said to have panic disorder.
Panic attack	An attack of overwhelming anxiety, fear, or terror is called panic attack.
Sympathetic	The sympathetic nervous system activates what is often termed the "fight or flight response". It is an automatic regulation system, that is, one that operates without the intervention of conscious thought.
Alprazolam	Alprazolam has a calming effect, with potential side effects of drowsiness, clumsiness, fatigue, and headache. The drug can also have more severe side effects, such as blurred vision, slurred speech, and changes in normal behavior. The drug is habituating, and users develop a tolerance.
Benzodiazepines	The benzodiazepines are a class of drugs with hypnotic, anxiolytic, anticonvulsant, amnestic and muscle relaxant properties. Benzodiazepines are often used for short-term relief of severe, disabling anxiety or insomnia.
Psychoactive	A psychoactive drug or psychotropic substance is a chemical that alters brain function,

drug	resulting in temporary changes in perception, mood, consciousness, or behavior. Such drugs are often used for recreational and spiritual purposes, as well as in medicine, especially for treating neurological and psychological illnesses.
Nucleus accumbens	A complex of neurons that is part of the brain's "pleasure pathway" responsible for the experience of reward is referred to as the nucleus accumbens.
Midbrain	Located between the hindbrain and forebrain, a region in which many nerve-fiber systems ascend and descend to connect the higher and lower portions of the brain is referred to as midbrain. It is archipallian in origin, meaning its general architecture is shared with the most ancient of vertebrates. Dopamine produced in the subtantia nigra plays a role in motivation and habituation of species from humans to the most elementary animals such as insects.
Opiates	A group of narcotics derived from the opium poppy that provide a euphoric rush and depress the nervous system are referred to as opiates.
Opioid	An opioid is any agent that binds to opioid receptors, found principally in the central nervous system and gastrointestinal tract.
Electrode	Any device used to electrically stimulate nerve tissue or to record its activity is an electrode.
Inference	Inference is the act or process of drawing a conclusion based solely on what one already knows.
Sensation	Sensation is the first stage in the chain of biochemical and neurologic events that begins with the impinging of a stimulus upon the receptor cells of a sensory organ, which then leads to perception, the mental state that is reflected in statements like "I see a uniformly blue wall."
Perception	Perception is the process of acquiring, interpreting, selecting, and organizing sensory information.
Noradrenaline	Noradrenaline is released from the adrenal glands as a hormone into the blood, but it is also a neurotransmitter in the nervous system. As a stress hormone, it affects parts of the human brain where attention and impulsivity are controlled. Along with epinephrine, this compound effects the fight-or-flight response, activating the sympathetic nervous system to directly increase heart rate, release energy from fat, and increase muscle readiness.
Cerebral cortex	The cerebral cortex is the outermost layer of the cerebrum and has a grey color. It is made up of four lobes and it is involved in many complex brain functions including memory, perceptual awareness, "thinking", language and consciousness. The cerebral cortex receives sensory information from many different sensory organs eg: eyes, ears, etc. and processes the information.
Activating effect	The arousal-producing effects of sex hormones that increase the likelihood of sexual behavior is called the activating effect.
Mental illness	Mental illness is the term formerly used to mean psychological disorder but less preferred because it implies that the causes of the disorder can be found in a medical disease process.
Survey	A method of scientific investigation in which a large sample of people answer questions about their attitudes or behavior is referred to as a survey.
Consciousness	The awareness of the sensations, thoughts, and feelings being experienced at a given moment is called consciousness.
Ethology	Where comparative psychology sees the study of animal behavior in the context of what is known about human psychology, ethology sees the study of animal behavior in the context of

what is known about animal anatomy and physiology.

Addiction

Addiction is an uncontrollable compulsion to repeat a behavior regardless of its consequences. Many drugs or behaviors can precipitate a pattern of conditions recognized as addiction, which include a craving for more of the drug or behavior, increased physiological tolerance to exposure, and withdrawal symptoms in the absence of the stimulus.

Drug addiction

Drug addiction, or substance dependence is the compulsive use of drugs, to the point where the user has no effective choice but to continue use.

Homeostasis

Homeostasis is the property of an open system, especially living organisms, to regulate its internal environment so as to maintain a stable condition, by means of multiple dynamic equilibrium adjustments controlled by interrelated regulation mechanisms.

Acquisition	Acquisition is the process of adapting to the environment, learning or becoming conditioned. In classical conditoning terms, it is the initial learning of the stimulus response link, which involves a neutral stimulus being associated with a unconditioned stimulus and becoming a conditioned stimulus.
Biological rhythm	A biological rhythm is a hypothetical cyclic pattern of alterations in physiology, emotions, and/or intellect
Metabolic rate	Metabolic rate refers to the rate at which the body burns calories to produce energy.
Species	Species refers to a reproductively isolated breeding population.
Maladaptive	In psychology, a behavior or trait is adaptive when it helps an individual adjust and function well within their social environment. A maladaptive behavior or trait is counterproductive to the individual.
Nocturnal	A person who exhibits nocturnal habits is referred to as a night owl.
Mental illness	Mental illness is the term formerly used to mean psychological disorder but less preferred because it implies that the causes of the disorder can be found in a medical disease process.
Stimulus	A change in an environmental condition that elicits a response is a stimulus.
Deprivation	Deprivation, is the loss or withholding of normal stimulation, nutrition, comfort, love, and so forth; a condition of lacking. The level of stimulation is less than what is required.
Perception	Perception is the process of acquiring, interpreting, selecting, and organizing sensory information.
Depression	In everyday language depression refers to any downturn in mood, which may be relatively transitory and perhaps due to something trivial. This is differentiated from Clinical depression which is marked by symptoms that last two weeks or more and are so severe that they interfere with daily living.
Circadian rhythm	The circadian rhythm is a name given to the "internal body clock" that regulates the (roughly) 24 hour cycle of biological processes in animals and plants.
Adaptation	Adaptation is a lowering of sensitivity to a stimulus following prolonged exposure to that stimulus. Behavioral adaptations are special ways a particular organism behaves to survive in its natural habitat.
Suprachiasmatic nucleus	The suprachiasmatic nucleus is in the hypothalamus and is so named because it resides immediately above the optic chaism. Its principal function is to create the circadian rhythm, which regulates the body functions over the 24-hour period.
Nucleus	In neuroanatomy, a cluster of cell bodies of neurons within the central nervous system is a nucleus.
Neuron	The neuron is the primary cell of the nervous system. They are found in the brain, the spinal cord, in the nerves and ganglia of the peripheral nervous system. It is a specialized cell that conducts impulses through the nervous system and contains three major parts: cell body, dendrites, and an axon. It can have many dendrites but only one axon.
Brain	The brain controls and coordinates most movement, behavior and homeostatic body functions such as heartbeat, blood pressure, fluid balance and body temperature. Functions of the brain are responsible for cognition, emotion, memory, motor learning and other sorts of learning. The brain is primarily made up of two types of cells: glia and neurons.
Lesion	A lesion is a non-specific term referring to abnormal tissue in the body. It can be caused by any disease process including trauma (physical, chemical, electrical), infection, neoplasm, metabolic and autoimmune.

Protein | A protein is a complex, high-molecular-weight organic compound that consists of amino acids joined by peptide bonds. It is essential to the structure and function of all living cells and viruses. Many are enzymes or subunits of enzymes.

Gene | A gene is an ultramicroscopic area of the chromosome. It is the smallest physical unit of the DNA molecule that carries a piece of hereditary information.

Photoreceptor | Photoreceptor cells are contained in the retina and are responsible for transducing, or converting, light into signals that can be ultimately transmitted to the brain via the optic nerve. A photoreceptor cell contains a membranous protein called an opsin which contains a pigment molecule called retinal.

Retina | The retina is a thin layer of cells at the back of the eyeball. It is the part of the eye which converts light into nervous signals. The retina contains photoreceptor cells which receive the light; the resulting neural signals then undergo complex processing by other neurons of the retina, and are transformed into action potentials in retinal ganglion cells whose axons form the optic nerve.

Hormone | A hormone is a chemical messenger from one cell (or group of cells) to another. The best known are those produced by endocrine glands, but they are produced by nearly every organ system. The function of hormones is to serve as a signal to the target cells; the action of the hormone is determined by the pattern of secretion and the signal transduction of the receiving tissue.

Gonads | The gonads are the organs that make gametes. Gametes are haploid germ cells. For example, sperm and egg cells are gametes. In the male the gonads are the testicles, and in the female the gonads are the ovaries.

Pineal gland | The pineal gland is a small endocrine gland. It is located near the center of the brain, between the two hemispheres and near the central switching point of the thalamic bodies. It is responsible for the production of melatonin, which has a role in regulating circadian rhythms.

Melatonin | Melatonin produced in the pineal gland acts as an endocrine hormone since it is released into the blood. Melatonin helps regulate sleep-wake or circadian rhythms.

Testes | Testes are the male reproductive glands or gonads; this is where sperm develop and are stored.

Hypothalamus | The hypothalamus is a region of the brain located below the thalamus, forming the major portion of the ventral region of the diencephalon and functioning to regulate certain metabolic processes and other autonomic activities.

Thalamus | An area near the center of the brain involved in the relay of sensory information to the cortex and in the functions of sleep and attention is the thalamus.

Sleep stages | Levels of sleep identified by brain-wave patterns and behavioral changes are called sleep stages.

Rem sleep | Sleep characterized by rapid eye movements, paralysis of large muscles, fast and irregular heart rate and respiration rate, increased brain-wave activity, and vivid dreams is referred to as REM sleep. An infant spends about half the time in REM sleep when sleeping.

Chronic | Chronic refers to a relatively long duration, usually more than a few months.

Sensation | Sensation is the first stage in the chain of biochemical and neurologic events that begins with the impinging of a stimulus upon the receptor cells of a sensory organ, which then leads to perception, the mental state that is reflected in statements like "I see a uniformly blue wall."

Syndrome — The term syndrome is the association of several clinically recognizable features, signs, symptoms, phenomena or characteristics which often occur together, so that the presence of one feature indicates the presence of the others.

Benzodiazepines — The benzodiazepines are a class of drugs with hypnotic, anxiolytic, anticonvulsant, amnestic and muscle relaxant properties. Benzodiazepines are often used for short-term relief of severe, disabling anxiety or insomnia.

Distal — Students can set both long-term (distal) and short-term (proximal) goals .

Variable — A variable refers to a measurable factor, characteristic, or attribute of an individual or a system.

Personality — Personality refers to the pattern of enduring characteristics that differentiates a person, the patterns of behaviors that make each individual unique.

Manifest content — In psychodynamic theory, the reported content of dreams is referred to as manifest content.

Latent content — In psychodynamic theory, the symbolized or underlying content of dreams is called latent content.

Psychoanalyst — A psychoanalyst is a specially trained therapist who attempts to treat the individual by uncovering and revealing to the individual otherwise subconscious factors that are contributing to some undesirable behavor.

Insight — Insight refers to a sudden awareness of the relationships among various elements that had previously appeared to be independent of one another.

Jung — Jung was in some aspects a response to Sigmund Freud's psychoanalysis. He proposed and developed the concepts of the extroverted and introverted personality, archetypes, and the collective unconscious. His work has been influential in psychiatry and in the study of religion, literature, and related fields.

Theories — Theories are logically self-consistent models or frameworks describing the behavior of a certain natural or social phenomenon. They are broad explanations and predictions concerning phenomena of interest.

Anxiety — Anxiety is a complex combination of the feeling of fear, apprehension and worry often accompanied by physical sensations such as palpitations, chest pain and/or shortness of breath.

Memory stores — In cognitive psychology, hypothetical constructs that are conceived of as places where information is held in the mind are memory stores.

Emotion — An emotion is a mental states that arise spontaneously, rather than through conscious effort. They are often accompanied by physiological changes.

Activation-synthesis — Activation-Synthesis is a neurobiological theory of dreams, put forward by Allan Hobson and Robert McCarley in 1977, which states that dreams are a random event caused by firing of neurons in the brain.

Hypothesis — A specific statement about behavior or mental processes that is testable through research is a hypothesis.

Stages — Stages represent relatively discrete periods of time in which functioning is qualitatively different from functioning at other periods.

Rapid Eye Movement — Rapid eye movement is the stage of sleep during which the most vivid (though not all) dreams occur. During this stage, the eyes move rapidly, and the activity of the brain's neurons is quite similar to that during waking hours. It is the lightest form of sleep in that people awakened during REM usually feel alert and refreshed.

Stage 4 sleep	The deepest stage of sleep, during which we are least responsive to outside stimulation is referred to as stage 4 sleep.
Adaptive behavior	An adaptive behavior increases the probability of the individual or organism to survive or exist within its environment.
Sleep patterns	The order and timing of daily sleep and waking periods are called sleep patterns.
Conservation	Conservation refers to the recognition that basic properties of substances such as weight and mass remain the same even when transformations merely alter their appearance.
Slow-wave sleep	Slow-wave sleep includes stages 3 and 4, during which low-frequency delta waves become prominent in EEG recordings.
Case study	A carefully drawn biography that may be obtained through interviews, questionnaires, and psychological tests is called a case study.
Sustained attention	Sustained attention is the state of readiness to detect and respond to small changes occurring at random times in the environment.
Microsleep	A microsleep is a period of sleep lasting a few seconds. It often occurs as a result of a sleep debt or mental fatigue.
Tricyclic antidepressant	A tricyclic antidepressant is of a class of antidepressant drugs first used in the 1950s. They are named after the drugs' molecular structure, which contains three rings of atoms.
Antidepressant	An antidepressant is a medication used primarily in the treatment of clinical depression. They are not thought to produce tolerance, although sudden withdrawal may produce adverse effects. They create little if any immediate change in mood and require between several days and several weeks to take effect.
Serotonin	Serotonin, a neurotransmitter, is believed to play an important part of the biochemistry of depression, bipolar disorder and anxiety. It is also believed to be influential on sexuality and appetite.
Correlation	A statistical technique for determining the degree of association between two or more variables is referred to as correlation.
Learning	Learning is a relatively permanent change in behavior that results from experience. Thus, to attribute a behavioral change to learning, the change must be relatively permanent and must result from experience.
Neuroscience	A field that combines the work of psychologists, biologists, biochemists, medical researchers, and others in the study of the structure and function of the nervous system is neuroscience.
Tranquilizer	A sedative, or tranquilizer, is a drug that depresses the central nervous system (CNS), which causes calmness, relaxation, reduction of anxiety, sleepiness, slowed breathing, slurred speech, staggering gait, poor judgment, and slow, uncertain reflexes.
Affect	A subjective feeling or emotional tone often accompanied by bodily expressions noticeable to others is called affect.
Reticular activating system	The reticular activating system is the part of the brain believed to be the center of arousal and motivation. It is situated between the brain stem and the central nervous system (CNS).
Electrode	Any device used to electrically stimulate nerve tissue or to record its activity is an electrode.
Axon	An axon, or "nerve fiber," is a long slender projection of a nerve cell, or "neuron," which conducts electrical impulses away from the neuron's cell body or soma.

Acute	Acute means sudden, sharp, and abrupt. Usually short in duration.
Acetylcholine	The chemical compound acetylcholine was the first neurotransmitter to be identified. It plays a role in learning, memory, and rapid eye movement sleep and causes the skeletal muscle fibers to contract.
Neocortex	The neocortex is part of the cerebral cortex which covers most of the surface of the cerebral hemispheres including the frontal, parietal, occipital, and temporal lobes. Often seen as the hallmark of human intelligence, the role of this structure in the brain appears to be involved in conscious thought, spatial reasoning, and sensory perception.
Narcolepsy	A serious sleep disorder characterized by excessive daytime sleepiness and sudden, uncontrollable attacks of REM sleep is called narcolepsy.
Insomnia	Insomnia is a sleep disorder characterized by an inability to sleep and/or to remain asleep for a reasonable period during the night.
Quantitative	A quantitative property is one that exists in a range of magnitudes, and can therefore be measured. Measurements of any particular quantitative property are expressed as as a specific quantity, referred to as a unit, multiplied by a number.
Schizophrenia	Schizophrenia is characterized by persistent defects in the perception or expression of reality. A person suffering from untreated schizophrenia typically demonstrates grossly disorganized thinking, and may also experience delusions or auditory hallucinations
Rebound insomnia	In a person with insomnia, the increase in sleep problems that can occur when medications are used to treat insomnia and then withdrawn is referred to as rebound insomnia.
Drug-dependency insomnia	Insomnia that follows withdrawal from sleeping pills is called drug-dependency insomnia.
Cataplexy	Sudden loss of muscle tone that accompanies narcolepsy is called cataplexy.
Hallucination	A hallucination is a sensory perception experienced in the absence of an external stimulus, as distinct from an illusion, which is a misperception of an external stimulus. They may occur in any sensory modality - visual, auditory, olfactory, gustatory, tactile, or mixed.
Motor neuron	A motor neuron is an efferent neuron that originates in the spinal cord and synapses with muscle fibers to facilitate muscle contraction and with muscle spindles to modify proprioceptive sensitivity.
Hallucinogenic	Certain drugs can affect the subjective qualities of perception, thought or emotion, resulting in altered interpretations of sensory input, alternate states of consciousness, or hallucinations. The term hallucinogenic is often broadly applied, especially in current scientific literature, to some or all of these substances.
Sleep apnea	Sleep apnea refers to a sleep disorder involving periods during sleep when breathing stops and the person must awaken briefly in order to breathe; major symptoms are excessive daytime sleepiness and loud snoring.
Sudden infant death syndrome	Sudden Infant Death Syndrome is the term for the sudden and unexplained death of an apparently healthy infant aged one month to one year.
Trait	An enduring personality characteristic that tends to lead to certain behaviors is called a trait. The term trait also means a genetically inherited feature of an organism.
Forebrain	The forebrain is the highest level of the brain. Key structures in the forebrain are the limbic system, thalamus, basal ganglia, hypothalamus, and cerebral cortex.
Amygdala	Located in the brain's medial temporal lobe, the almond-shaped amygdala is believed to play a key role in the emotions. It forms part of the limbic system and is linked to both fear

responses and pleasure. Its size is positively correlated with aggressive behavior across species.

Ritalin — Ritalin, a methylphenidate, is a central nervous system stimulant. It has a "calming" effect on many children who have ADHD, reducing impulsive behavior and the tendency to "act out", and helps them concentrate on schoolwork and other tasks.

Consciousness — The awareness of the sensations, thoughts, and feelings being experienced at a given moment is called consciousness.

Sedative — A sedative is a drug that depresses the central nervous system (CNS), which causes calmness, relaxation, reduction of anxiety, sleepiness, slowed breathing, slurred speech, staggering gait, poor judgment, and slow, uncertain reflexes.

Research method — The scope of the research method is to produce some new knowledge. This, in principle, can take three main forms: Exploratory research; Constructive research; and Empirical research.

Reticular formation — Reticular formation is a part of the brain which is involved in stereotypical actions, such as walking, sleeping, and lying down. The reticular formation, phylogenetically one of the oldest portions of the brain, is a poorly-differentiated area of the brain stem.

Ethology — Where comparative psychology sees the study of animal behavior in the context of what is known about human psychology, ethology sees the study of animal behavior in the context of what is known about animal anatomy and physiology.

Consciousness	The awareness of the sensations, thoughts, and feelings being experienced at a given moment is called consciousness.
Emotion	An emotion is a mental states that arise spontaneously, rather than through conscious effort. They are often accompanied by physiological changes.
Plasticity	The capacity for modification and change is referred to as plasticity.
Brain	The brain controls and coordinates most movement, behavior and homeostatic body functions such as heartbeat, blood pressure, fluid balance and body temperature. Functions of the brain are responsible for cognition, emotion, memory, motor learning and other sorts of learning. The brain is primarily made up of two types of cells: glia and neurons.
Puberty	Puberty refers to the process of physical changes by which a child's body becomes an adult body capable of reproduction.
Neuroscience	A field that combines the work of psychologists, biologists, biochemists, medical researchers, and others in the study of the structure and function of the nervous system is neuroscience.
Learning	Learning is a relatively permanent change in behavior that results from experience. Thus, to attribute a behavioral change to learning, the change must be relatively permanent and must result from experience.
Brain circuits	Neurotransmitter currents or neural pathways in the brain are referred to as brain circuits.
Synapse	A synapse is specialized junction through which cells of the nervous system signal to one another and to non-neuronal cells such as muscles or glands.
Neuron	The neuron is the primary cell of the nervous system. They are found in the brain, the spinal cord, in the nerves and ganglia of the peripheral nervous system. It is a specialized cell that conducts impulses through the nervous system and contains three major parts: cell body, dendrites, and an axon. It can have many dendrites but only one axon.
Addiction	Addiction is an uncontrollable compulsion to repeat a behavior regardless of its consequences. Many drugs or behaviors can precipitate a pattern of conditions recognized as addiction, which include a craving for more of the drug or behavior, increased physiological tolerance to exposure, and withdrawal symptoms in the absence of the stimulus.
Hormone	A hormone is a chemical messenger from one cell (or group of cells) to another. The best known are those produced by endocrine glands, but they are produced by nearly every organ system. The function of hormones is to serve as a signal to the target cells; the action of the hormone is determined by the pattern of secretion and the signal transduction of the receiving tissue.
Mental Representation	Stage six of the sensorimotor substages, Mental representation, 18 months to 2 years, marks the beginnings of insight, or true creativity. This marks the passage into unique thought in Piaget's later three areas of development.
Species	Species refers to a reproductively isolated breeding population.
Pavlovian conditioning	Pavlovian conditioning, synonymous with classical conditioning is a type of learning found in animals, caused by the association (or pairing) of two stimuli or what Ivan Pavlov described as the learning of conditional behavior, therefore called conditioning.
Stimulus	A change in an environmental condition that elicits a response is a stimulus.
Pavlov	Pavlov first described the phenomenon now known as classical conditioning in experiments with dogs.
Conditioning	Conditioning describes the process by which behaviors can be learned or modified through

	interaction with the environment.
Unconditioned response	An Unconditioned Response is the response elicited to an unconditioned stimulus. It is a natural, automatic response.
Conditioned stimulus	A previously neutral stimulus that elicits the conditioned response because of being repeatedly paired with a stimulus that naturally elicited that response, is called a conditioned stimulus.
Cerebellum	The cerebellum is located in the inferior posterior portion of the head (the hindbrain), directly dorsal to the brainstem and pons, inferior to the occipital lobe. The cerebellum is a region of the brain that plays an important role in the integration of sensory perception and fine motor output.
Predisposition	Predisposition refers to an inclination or diathesis to respond in a certain way, either inborn or acquired. In abnormal psychology, it is a factor that lowers the ability to withstand stress and inclines the individual toward pathology.
Respondent Conditioning	Respondent conditioning refers to behavior that is elicited involuntarily as a reaction to a stimulus. Respondent behavior is identical to classical conditioning UC to UR relationships.
Classical conditioning	Classical conditioning is a simple form of learning in which an organism comes to associate or anticipate events. A neutral stimulus comes to evoke the response usually evoked by a natural or unconditioned stimulus by being paired repeatedly with the unconditioned stimulus.
Neutral stimulus	A stimulus prior to conditioning that does not naturally result in the response of interest is called a neutral stimulus.
Conditioned emotional response	Conditioned emotional response refers to an emotional response that has been linked to a previously non-emotional stimulus by classical conditioning.
Amygdala	Located in the brain's medial temporal lobe, the almond-shaped amygdala is believed to play a key role in the emotions. It forms part of the limbic system and is linked to both fear responses and pleasure. Its size is positively correlated with aggressive behavior across species.
Instrumental conditioning	Operant conditioning, sometimes called instrumental conditioning, was first extensively studied by Thorndike. In instrumental conditioning, the organism must act in a certain way before it is reinforced; that is, reinforcement is contingent on the organism's behavior.
Thorndike	Thorndike worked in animal behavior and the learning process leading to the theory of connectionism. Among his most famous contributions were his research on cats escaping from puzzle boxes, and his formulation of the Law of Effect.
Skinner	Skinner conducted research on shaping behavior through positive and negative reinforcement, and demonstrated operant conditioning, a technique which he developed in contrast with classical conditioning.
Puzzle Box	The puzzle box experiments were motivated in part by Thorndike's dislike for statements that animals made use of extraordinary factulties such as insight in their problem solving. He reasoned that if the animals were showing insight, then their time to escape would suddenly drop to a negligible period, which would also be shown in the learning curve as an abrupt drop; while animals using a more ordinary method of trial and error would show gradual curves. His finding was that cats consistently showed gradual learning. He asserted that the connection between the box and the motions the cat used to escape was strengthened by each escape.
Distal	Students can set both long-term (distal) and short-term (proximal) goals .

Implicit memory	Implicit memory is the long-term memory of skills and procedures, or "how to" knowledge. It is often not easily verbalized, but can be used without consciously thinking about it.
Amnesia	Amnesia is a condition in which memory is disturbed. The causes of amnesia are organic or functional. Organic causes include damage to the brain, through trauma or disease, or use of certain (generally sedative) drugs.
Control subjects	Control subjects are participants in an experiment who do not receive the treatment effect but for whom all other conditions are held comparable to those of experimental subjects.
Explicit memory	Intentional or conscious recollection of information is referred to as explicit memory. Children under age three are usually poorest at explicit memory which may be due to the immaturity of the prefrontal lobes of the brain, which are believed to play an important role in memory for events.
Procedural memory	Procedural memory is the long-term memory of skills and procedures, or "how to" knowledge.
Dichotomy	A dichotomy is the division of a proposition into two parts which are both mutually exclusive – i.e. both cannot be simultaneously true – and jointly exhaustive – i.e. they cover the full range of possible outcomes. They are often contrasting and spoken of as "opposites".
Sensory receptor	A sensory receptor is a structure that recognizes a stimulus in the environment of an organism. In response to stimuli the sensory receptor initiates sensory transduction by creating graded potentials or action potentials in the same cell or in an adjacent one.
Top-down	In the Top-down model an overview of the system is formulated, without going into detail for any part of it. Each part of the system is then refined by designing it in more detail. Each new part may then be refined again, defining it in yet more detail until the entire specification is detailed enough to validate the Model.
Encoding	Encoding refers to interpreting; transforming; modifying information so that it can be placed in memory. It is the first stage of information processing.
Loftus	Loftus works on human memory and how it can be changed by facts, ideas, suggestions and other forms of post-event information. One of her famous studies include the "car accident" study, which was an example of the misinformation effect.
Pathology	Pathology is the study of the processes underlying disease and other forms of illness, harmful abnormality, or dysfunction.
Forebrain	The forebrain is the highest level of the brain. Key structures in the forebrain are the limbic system, thalamus, basal ganglia, hypothalamus, and cerebral cortex.
Homogeneous	In biology homogeneous has a meaning similar to its meaning in mathematics. Generally it means "the same" or "of the same quality or general property".
Protein	A protein is a complex, high-molecular-weight organic compound that consists of amino acids joined by peptide bonds. It is essential to the structure and function of all living cells and viruses. Many are enzymes or subunits of enzymes.
Neuritic plaque	Neuritic plaque is a cluster of dead neurons often found in the brains of people with Alzheimer's disease.
Plaques	Plaques refer to small, round areas composed of remnants of lost neurons and beta-amyloid, a waxy protein deposit; present in the brains of patients with Alzheimer's disease.
Lesion	A lesion is a non-specific term referring to abnormal tissue in the body. It can be caused by any disease process including trauma (physical, chemical, electrical), infection, neoplasm, metabolic and autoimmune.

Hippocampus	The hippocampus is a part of the brain located inside the temporal lobe. It forms a part of the limbic system and plays a part in memory and navigation.
Hypothesis	A specific statement about behavior or mental processes that is testable through research is a hypothesis.
Anterograde amnesia	Anterograde amnesia is a form of amnesia where new events are not transferred to long-term memory, so the sufferer will not be able to remember anything that occurs after the onset of this type of amnesia for more than a few moments.
Malnutrition	Malnutrition is a general term for the medical condition in a person or animal caused by an unbalanced diet—either too little or too much food, or a diet missing one or more important nutrients.
Alcoholism	A disorder that involves long-term, repeated, uncontrolled, compulsive, and excessive use of alcoholic beverages and that impairs the drinker's health and work and social relationships is called alcoholism.
Chronic	Chronic refers to a relatively long duration, usually more than a few months.
Magnetic resonance imaging	Magnetic resonance imaging is a method of creating images of the inside of opaque organs in living organisms as well as detecting the amount of bound water in geological structures. It is primarily used to demonstrate pathological or other physiological alterations of living tissues and is a commonly used form of medical imaging.
Temporal lobe	The temporal lobe is part of the cerebrum. It lies at the side of the brain, beneath the lateral or Sylvian fissure. Adjacent areas in the superior, posterior and lateral parts of the temporal lobe are involved in high-level auditory processing.
Frontal lobe	The frontal lobe comprises four major folds of cortical tissue: the precentral gyrus, superior gyrus and the middle gyrus of the frontal gyri, the inferior frontal gyrus. It has been found to play a part in impulse control, judgement, language, memory, motor function, problem solving, sexual behavior, socialization and spontaneity.
Short-term memory	Short-term memory is that part of memory which stores a limited amount of information for a limited amount of time (roughly 30-45 seconds). The second key concept associated with a short-term memory is that it has a finite capacity.
Prefrontal cortex	The prefrontal cortex is the anterior part of the frontal lobes of the brain, lying in front of the motor and associative areas. It has been implicated in planning complex cognitive behaviors, personality expression and moderating correct social behavior. The prefrontal cortex continues to develop until around age 6.
Syndrome	The term syndrome is the association of several clinically recognizable features, signs, symptoms, phenomena or characteristics which often occur together, so that the presence of one feature indicates the presence of the others.
Diencephalon	The diencephalon is the region of the brain that includes the epithalamus, thalamus, and hypothalamus. It is located above the mesencephalon of the brain stem. Sensory information is relayed between the brain stem and the rest of the brain regions
Hypothalamus	The hypothalamus is a region of the brain located below the thalamus, forming the major portion of the ventral region of the diencephalon and functioning to regulate certain metabolic processes and other autonomic activities.
Thalamus	An area near the center of the brain involved in the relay of sensory information to the cortex and in the functions of sleep and attention is the thalamus.
Cerebral cortex	The cerebral cortex is the outermost layer of the cerebrum and has a grey color. It is made up of four lobes and it is involved in many complex brain functions including memory,

	perceptual awareness, "thinking", language and consciousness. The cerebral cortex receives sensory information from many different sensory organs eg: eyes, ears, etc. and processes the information.
Basal ganglia	The basal ganglia are a group of nuclei in the brain associated with motor and learning functions.
Projection	Attributing one's own undesirable thoughts, impulses, traits, or behaviors to others is referred to as projection.
Substantia nigra	The substantia nigra is a portion of the midbrain thought to be involved in certain aspects of movement and attention. Degeneration of cells in this region is the principle pathology that underlies Parkinson's disease.
Dopamine	Dopamine is critical to the way the brain controls our movements and is a crucial part of the basal ganglia motor loop. It is commonly associated with the 'pleasure system' of the brain, providing feelings of enjoyment and reinforcement to motivate us to do, or continue doing, certain activities.
Fear response	In the Mowrer-Miller theory, a response to a threatening or noxious situation that is covert but that is assumed to function as a stimulus to produce measurable physiological changes in the body and observable overt behavior is referred to as the fear response.
Operant Conditioning	A simple form of learning in which an organism learns to engage in behavior because it is reinforced is referred to as operant conditioning. The consequences of a behavior produce changes in the probability of the behavior's occurence.
Perception	Perception is the process of acquiring, interpreting, selecting, and organizing sensory information.
Gyrus	A gyrus is a ridge on the cerebral cortex. It is generally surrounded by one or more sulci.
Long-term memory	Long-term memory is memory that lasts from over 30 seconds to years.
Nervous system	The body's electrochemical communication circuitry, made up of billions of neurons is a nervous system.
Parietal lobe	The parietal lobe is positioned above (superior to) the occipital lobe and behind (posterior to) the frontal lobe. It plays important roles in integrating sensory information from various senses, and in the manipulation of objects.
Memory trace	A memory trace is the postulated assumed change in the nervous system that reflects the impression made by a stimulus. They are said to be held in sensory registers.
Lashley	Lashley failed to find a single biological locus of memory suggesting to him that memories were not localized to one part of the brain, but were widely distributed throughout the cortex.
Working hypothesis	A simple statement of what is expected to happen in an experiment is a working hypothesis.
Epilepsy	Epilepsy is a chronic neurological condition characterized by recurrent unprovoked neural discharges. It is commonly controlled with medication, although surgical methods are used as well.
Penfield	Penfield treated patients with severe epilepsy by destroying nerve cells in the brain. Before operating, he stimulated the brain with electrical probes while the patients were conscious on the operating table, and observed their responses. It allowed him to create maps of sensory and motor functions.
Seizure	A seizure is a temporary alteration in brain function expressed as a changed mental state,

	tonic or clonic movements and various other symptoms. They are due to temporary abnormal electrical activity of a group of brain cells.
Wechsler Adult Intelligence Scale	Wechsler adult intelligence scale is an individual intelligence test for adults that yields separate verbal and performance IQ scores as well as an overall IQ score.
Acute	Acute means sudden, sharp, and abrupt. Usually short in duration.
Encephalitis	Encephalitis is an acute inflammation of the brain, commonly caused by a viral infection.
Affect	A subjective feeling or emotional tone often accompanied by bodily expressions noticeable to others is called affect.
Anatomy	Anatomy is the branch of biology that deals with the structure and organization of living things. It can be divided into animal anatomy (zootomy) and plant anatomy (phytonomy). Major branches of anatomy include comparative anatomy, histology, and human anatomy.
Psychological test	Psychological test refers to a standardized measure of a sample of a person's behavior.
Attention	Attention is the cognitive process of selectively concentrating on one thing while ignoring other things. Psychologists have labeled three types of attention: sustained attention, selective attention, and divided attention.
Retrograde amnesia	Retrograde amnesia is a form of amnesia where someone will be unable to recall events that occurred before the onset of amnesia.
Insight	Insight refers to a sudden awareness of the relationships among various elements that had previously appeared to be independent of one another.
Thiamine	Thiamine, also known as vitamin B1, is a colorless compound with chemical formula $C_{12}H_{17}ClN_4OS$. Systemic thiamine deficiency can lead to myriad problems including neurodegeneration, wasting, and death. Well-known syndromes caused by lack of thiamine due to malnutrition or a diet high in thiaminase-rich foods include Wernicke-Korsakoff syndrome and beriberi, diseases also common in chronic abusers of alcohol.
Prognosis	A forecast about the probable course of an illess is referred to as prognosis.
Feedback	Feedback refers to information returned to a person about the effects a response has had.
Stroke	A stroke occurs when the blood supply to a part of the brain is suddenly interrupted by occlusion, by hemorrhage, or other causes
Neocortex	The neocortex is part of the cerebral cortex which covers most of the surface of the cerebral hemispheres including the frontal, parietal, occipital, and temporal lobes. Often seen as the hallmark of human intelligence, the role of this structure in the brain appears to be involved in conscious thought, spatial reasoning, and sensory perception.
Cajal	Cajal, Nobel laureate, is considered to be the father of modern neuroscience. He postulated that the nervous system is made up of billions of separate nerve cells (neurons) and that nerve cells are polarized. He suggested that neurons communicate with each other via specialized junctions. This hypothesis became the basis of the neuron doctrine.
Nerve	A nerve is an enclosed, cable-like bundle of nerve fibers or axons, which includes the glia that ensheath the axons in myelin. Neurons are sometimes called nerve cells, though this term is technically imprecise since many neurons do not form nerves.
Dendrite	A dendrite is a slender, typically branched projection of a nerve cell, or "neuron," which conducts the electrical stimulation received from other cells to the body or soma of the cell from which it projects. This stimulation arrives through synapses, which typically are

	located near the tips of the dendrites and away from the soma.
Morphology	Morphology is a subdiscipline of linguistics that studies word structure. While words are generally accepted as being the smallest units of syntax, it is clear that in most (if not all) languages, words can be related to other words by rules. .
Dendritic spine	A dendritic spine is a small membranous extrusion that protrudes from a dendrite and forms one half of a synapse. Changes in dendritic spine density underlie many brain functions, including motivation, learning, and memory. In particular, long-term memory is mediated in part by the growth of new dendritic spines to reinforce a particular neural pathway.
Axon	An axon, or "nerve fiber," is a long slender projection of a nerve cell, or "neuron," which conducts electrical impulses away from the neuron's cell body or soma.
Axon terminal	A swelling at the end of an axon that is designed to release a chemical substance onto another neuron, muscle cell, or gland cell is called the axon terminal.
Olfactory bulb	The olfactory bulb is a part of the brain that is a distinct outgrowth from the forebrain. It plays a major role in olfaction, which is the perception of smells. The olfactory bulb receives direct input from olfactory nerve, made up of the axons from approximately 10 million olfactory receptor neurons in the olfactory mucosa, a region of the nasal cavity.
Hebb	Hebb demonstrated that the rearing of rats in an enriched environment could alter neural development and that sensory - neural connections were shaped by experience. He is famous for developing the concept of neural nets. He also believed that learning early in life is of the incremental variety, whereas later it is cognitive, insightful, and more all-or-none.
Enrichment	Deliberately making an environment more novel, complex, and perceptually or intellectually stimulating is referred to as enrichment.
Visual cortex	The visual cortex is the general term applied to both the primary visual cortex and the visual association area. Anatomically, the visual cortex occupies the entire occipital lobe, the inferior temporal lobe (IT), posterior parts of the parietal lobe, and a few small regions in the frontal lobe.
Tactile	Pertaining to the sense of touch is referred to as tactile.
Microelectrode	An electrical wire so small that it can be used either to monitor the electrical activity of a single neuron or to stimulate activity within it is a microelectrode.
Motor cortex	Motor cortex refers to the section of cortex that lies in the frontal lobe, just across the central fissure from the sensory cortex. Neural impulses in the motor cortex are linked to muscular responses throughout the body.
Functional magnetic resonance imaging	Functional Magnetic Resonance Imaging describes the use of MRI to measure hemodynamic signals related to neural activity in the brain or spinal cord of humans or other animals. It is one of the most recently developed forms of brain imaging.
Sensory nerves	Sensory nerves bring impulses toward the central nervous system.
Somatosensory	Somatosensory system consists of the various sensory receptors that trigger the experiences labelled as touch or pressure, temperature, pain, and the sensations of muscle movement and joint position including posture, movement, and facial expression.
Acquisition	Acquisition is the process of adapting to the environment, learning or becoming conditioned. In classical conditoning terms, it is the initial learning of the stimulus response link, which involves a neutral stimulus being associated with a unconditioned stimulus and becoming a conditioned stimulus.
Retrieval	Retrieval is the location of stored information and its subsequent return to consciousness.

	It is the third stage of information processing.
Receptive field	The receptive field of a sensory neuron is a region of sensitivity in which the presence of a stimulus will alter the firing of that neuron.
Individual differences	Individual differences psychology studies the ways in which individual people differ in their behavior. This is distinguished from other aspects of psychology in that although psychology is ostensibly a study of individuals, modern psychologists invariably study groups.
Estrogen	Estrogen is a group of steroid compounds that function as the primary female sex hormone. They are produced primarily by developing follicles in the ovaries, the corpus luteum and the placenta.
Pituitary gland	The pituitary gland is an endocrine gland about the size of a pea that sits in the small, bony cavity at the base of the brain. The pituitary gland secretes hormones regulating a wide variety of bodily activities, including trophic hormones that stimulate other endocrine glands.
Adrenal cortex	Adrenal cortex refers to the outer layer of the adrenal glands, which produces hormones that affect salt intake, reactions to stress, and sexual development.
Metabolism	Metabolism is the biochemical modification of chemical compounds in living organisms and cells.
Amphetamine	Amphetamine is a synthetic stimulant used to suppress the appetite, control weight, and treat disorders including narcolepsy and ADHD. It is also used recreationally and for performance enhancement.
Stimulant	A stimulant is a drug which increases the activity of the sympathetic nervous system and produces a sense of euphoria or awakeness.
Nicotine	Nicotine is an organic compound, an alkaloid found naturally throughout the tobacco plant, with a high concentration in the leaves. It is a potent nerve poison and is included in many insecticides. In lower concentrations, the substance is a stimulant and is one of the main factors leading to the pleasure and habit-forming qualities of tobacco smoking.
Cocaine	Cocaine is a crystalline tropane alkaloid that is obtained from the leaves of the coca plant. It is a stimulant of the central nervous system and an appetite suppressant, creating what has been described as a euphoric sense of happiness and increased energy.
Receptor	A sensory receptor is a structure that recognizes a stimulus in the internal or external environment of an organism. In response to stimuli the sensory receptor initiates sensory transduction by creating graded potentials or action potentials in the same cell or in an adjacent one.
Psychoactive drug	A psychoactive drug or psychotropic substance is a chemical that alters brain function, resulting in temporary changes in perception, mood, consciousness, or behavior. Such drugs are often used for recreational and spiritual purposes, as well as in medicine, especially for treating neurological and psychological illnesses.
Sensitization	Sensitization is a process whereby an organism is made more responsive to certain aspects of its environment. For example, increases in the effects of a drug as a result of repeated administration. Also known as reverse tolerance.
Morphine	Morphine, the principal active agent in opium, is a powerful opioid analgesic drug. According to recent research, it may also be produced naturally by the human brain. Morphine is usually highly addictive, and tolerance and physical and psychological dependence develop quickly.
Nucleus accumbens	A complex of neurons that is part of the brain's "pleasure pathway" responsible for the experience of reward is referred to as the nucleus accumbens.

Relearning	Relearning refers to a measure of retention used in experiments on memory. Material is usually relearned more quickly than it is learned initially.
Compensation	In personaility, compensation, according to Adler, is an effort to overcome imagined or real inferiorities by developing one's abilities.
Behavioral therapy	The treatment of a mental disorder through the application of basic principles of conditioning and learning is called behavioral therapy.
Liver	The liver plays a major role in metabolism and has a number of functions in the body including detoxification, glycogen storage and plasma protein synthesis. It also produces bile, which is important for digestion. The liver converts most carbohydrates, proteing, and fats into glucose.
Neurogenesis	Neurogenesis literally means "birth of neurons". Neurogenesis is most prevalent during pre-natal development and is the process by which neurons are created to populate the growing brain.
Central nervous system	The vertebrate central nervous system consists of the brain and spinal cord.
Somatosensory cortex	The primary somatosensory cortex is across the central sulcus and behind the primary motor cortex configured to generally correspond with the arrangement of nearby motor cells related to specific body parts. It is the main sensory receptive area for the sense of touch.

Neocortex	The neocortex is part of the cerebral cortex which covers most of the surface of the cerebral hemispheres including the frontal, parietal, occipital, and temporal lobes. Often seen as the hallmark of human intelligence, the role of this structure in the brain appears to be involved in conscious thought, spatial reasoning, and sensory perception.
Cell assembly	Hebb described the interlinking of neurons as creating what he called the cell assembly, a diffuse structure comprising cells in the cortex and diencephalon, capable of acting briefly as a closed system. That functional neural nets can be assembled by experience is one of the more important ideas in neuropsychological theory of the twentieth century.
Hebb	Hebb demonstrated that the rearing of rats in an enriched environment could alter neural development and that sensory - neural connections were shaped by experience. He is famous for developing the concept of neural nets. He also believed that learning early in life is of the incremental variety, whereas later it is cognitive, insightful, and more all-or-none.
Neuron	The neuron is the primary cell of the nervous system. They are found in the brain, the spinal cord, in the nerves and ganglia of the peripheral nervous system. It is a specialized cell that conducts impulses through the nervous system and contains three major parts: cell body, dendrites, and an axon. It can have many dendrites but only one axon.
Construct	A generalized concept, such as anxiety or gravity, is a construct.
Brain	The brain controls and coordinates most movement, behavior and homeostatic body functions such as heartbeat, blood pressure, fluid balance and body temperature. Functions of the brain are responsible for cognition, emotion, memory, motor learning and other sorts of learning. The brain is primarily made up of two types of cells: glia and neurons.
Ventral stream	The primate visual system consists of about thirty areas of the cerebral cortex called the visual cortex. The visual cortex is divided into the ventral stream and the dorsal stream. The ventral stream is associated with object recognition and form representation.
Cerebral hemisphere	Either of the two halves that make up the cerebrum is referred to as a cerebral hemisphere. The hemispheres operate together, linked by the corpus callosum, a very large bundle of nerve fibers, and also by other smaller commissures.
Individual differences	Individual differences psychology studies the ways in which individual people differ in their behavior. This is distinguished from other aspects of psychology in that although psychology is ostensibly a study of individuals, modern psychologists invariably study groups.
Motivation	In psychology, motivation is the driving force (desire) behind all actions of an organism.
Emotion	An emotion is a mental states that arise spontaneously, rather than through conscious effort. They are often accompanied by physiological changes.
Cognition	The intellectual processes through which information is obtained, transformed, stored, retrieved, and otherwise used is cognition.
Sensory receptor	A sensory receptor is a structure that recognizes a stimulus in the environment of an organism. In response to stimuli the sensory receptor initiates sensory transduction by creating graded potentials or action potentials in the same cell or in an adjacent one.
Autonomic nervous system	A division of the peripheral nervous system, the autonomic nervous system, regulates glands and activities such as heartbeat, respiration, digestion, and dilation of the pupils. It is responsible for homeostasis, maintaining a relatively constant internal environment .
Syntax	Syntax can be described as the study of the rules, or "patterned relations" that govern the way the words in a sentence come together.
Universal grammar	Chomsky's theory that all the world's languages share a similar underlying structure is called universal grammar. It attempts to explain language acquisition in general, not

	describe specific languages.
Perception	Perception is the process of acquiring, interpreting, selecting, and organizing sensory information.
Feedback	Feedback refers to information returned to a person about the effects a response has had.
Predisposition	Predisposition refers to an inclination or diathesis to respond in a certain way, either inborn or acquired. In abnormal psychology, it is a factor that lowers the ability to withstand stress and inclines the individual toward pathology.
Frontal lobe	The frontal lobe comprises four major folds of cortical tissue: the precentral gyrus, superior gyrus and the middle gyrus of the frontal gyri, the inferior frontal gyrus. It has been found to play a part in impulse control, judgement, language, memory, motor function, problem solving, sexual behavior, socialization and spontaneity.
Apparent motion	Apparent motion is the perceived motion of an object when all that has been presented to the eyes is one or a series of stills.
Action potential	The electrical impulse that provides the basis for the conduction of a neural impulse along an axon of a neuron is the action potential. When a biological cell or patch of membrane undergoes an action potential, or electrical excitation, the polarity of the transmembrane voltage swings rapidly from negative to positive and back.
Correlation	A statistical technique for determining the degree of association between two or more variables is referred to as correlation.
Threshold	In general, a threshold is a fixed location or value where an abrupt change is observed. In the sensory modalities, it is the minimum amount of stimulus energy necessary to elicit a sensory response.
Neural network	A clusters of neurons that is interconnected to process information is referred to as a neural network.
Association cortex	Region of the cerebral cortex in which the highest intellectual functions, including thinking and problem solving, occur is the association cortex.
Parietal lobe	The parietal lobe is positioned above (superior to) the occipital lobe and behind (posterior to) the frontal lobe. It plays important roles in integrating sensory information from various senses, and in the manipulation of objects.
Primary motor area	The primary motor area is a group of networked cells that controls movements of specific body parts associated with cell groups in that area of the brain. The area is closely linked by neural networks to corresponding areas in the primary somatosensory cortex.
Left hemisphere	The left hemisphere of the cortex controls the right side of the body, coordinates complex movements, and, in 95% of people, controls the production of speech and written language.
Temporal lobe	The temporal lobe is part of the cerebrum. It lies at the side of the brain, beneath the lateral or Sylvian fissure. Adjacent areas in the superior, posterior and lateral parts of the temporal lobe are involved in high-level auditory processing.
Senses	The senses are systems that consist of a sensory cell type that respond to a specific kind of physical energy, and that correspond to a defined region within the brain where the signals are received and interpreted.
Motor cortex	Motor cortex refers to the section of cortex that lies in the frontal lobe, just across the central fissure from the sensory cortex. Neural impulses in the motor cortex are linked to muscular responses throughout the body.
Thalamus	An area near the center of the brain involved in the relay of sensory information to the

	cortex and in the functions of sleep and attention is the thalamus.
Lesion	A lesion is a non-specific term referring to abnormal tissue in the body. It can be caused by any disease process including trauma (physical, chemical, electrical), infection, neoplasm, metabolic and autoimmune.
Evolution	Commonly used to refer to gradual change, evolution is the change in the frequency of alleles within a population from one generation to the next. This change may be caused by different mechanisms, including natural selection, genetic drift, or changes in population (gene flow).
Mental Representation	Stage six of the sensorimotor substages, Mental representation, 18 months to 2 years, marks the beginnings of insight, or true creativity. This marks the passage into unique thought in Piaget's later three areas of development.
Spatial cognition	Spatial cognition involves processing visual information in terms of spatial relationships, including spatial orientation, spatial visualization, and object and location memory.
Spatial orientation	Spatial orientation refers to how people understand the placement of objects in space with themselves as the reference point.
Attention	Attention is the cognitive process of selectively concentrating on one thing while ignoring other things. Psychologists have labeled three types of attention: sustained attention, selective attention, and divided attention.
William James	Functionalism as a psychology developed out of Pragmatism as a philosophy: To find the meaning of an idea, you have to look at its consequences. This led William James and his students towards an emphasis on cause and effect, prediction and control, and observation of environment and behavior, over the careful introspection of the Structuralists.
Mental processes	The thoughts, feelings, and motives that each of us experiences privately but that cannot be observed directly are called mental processes.
Stimulus	A change in an environmental condition that elicits a response is a stimulus.
Fixation	Fixation in abnormal psychology is the state where an individual becomes obsessed with an attachment to another human, animal or inanimate object. Fixation in vision refers to maintaining the gaze in a constant direction. .
Receptive field	The receptive field of a sensory neuron is a region of sensitivity in which the presence of a stimulus will alter the firing of that neuron.
Variable	A variable refers to a measurable factor, characteristic, or attribute of an individual or a system.
Affect	A subjective feeling or emotional tone often accompanied by bodily expressions noticeable to others is called affect.
Baseline	Measure of a particular behavior or process taken before the introduction of the independent variable or treatment is called the baseline.
Right hemisphere	The brain is divided into left and right cerebral hemispheres. The right hemisphere of the cortex controls the left side of the body.
Extinction	In operant extinction, if no reinforcement is delivered after the response, gradually the behavior will no longer occur in the presence of the stimulus. The process is more rapid following continuous reinforcement rather than after partial reinforcement. In Classical Conditioning, repeated presentations of the CS without being followed by the US results in the extinction of the CS.
Association areas	Association areas refer to the site of the higher mental processes such as thought, language, memory, and speech.

Hippocampus	The hippocampus is a part of the brain located inside the temporal lobe. It forms a part of the limbic system and plays a part in memory and navigation.
Neuroscience	A field that combines the work of psychologists, biologists, biochemists, medical researchers, and others in the study of the structure and function of the nervous system is neuroscience.
Paul Broca	Paul Broca became famous with his discovery of the speech center, the third circumvolution of the frontal lobe. He arrived at this discovery by studying the brains of aphasic patients (persons unable to talk).
Neuropsychology	Neuropsychology is a branch of psychology that aims to understand how the structure and function of the brain relates to specific psychological processes.
Stroke	A stroke occurs when the blood supply to a part of the brain is suddenly interrupted by occlusion, by hemorrhage, or other causes
Tumor	A tumor is an abnormal growth that when located in the brain can either be malignant and directly destroy brain tissue, or be benign and disrupt functioning by increasing intracranial pressure.
Neuropsychological test	A neuropsychological test use specifically designed tasks used to measure a psychological function known to be linked to a particular brain structure or pathway. They usually involve the systematic administration of clearly defined procedures in a formal environment.
Reasoning	Reasoning is the act of using reason to derive a conclusion from certain premises. There are two main methods to reach a conclusion,deductive reasoning and inductive reasoning.
Psychological test	Psychological test refers to a standardized measure of a sample of a person's behavior.
Magnetic resonance imaging	Magnetic resonance imaging is a method of creating images of the inside of opaque organs in living organisms as well as detecting the amount of bound water in geological structures. It is primarily used to demonstrate pathological or other physiological alterations of living tissues and is a commonly used form of medical imaging.
Research method	The scope of the research method is to produce some new knowledge. This, in principle, can take three main forms: Exploratory research; Constructive research; and Empirical research.
Visual cortex	The visual cortex is the general term applied to both the primary visual cortex and the visual association area. Anatomically, the visual cortex occupies the entire occipital lobe, the inferior temporal lobe (IT), posterior parts of the parietal lobe, and a few small regions in the frontal lobe.
Occipital lobe	The occipital lobe is the smallest of four true lobes in the human brain. Located in the rearmost portion of the skull, the occipital lobe is part of the forebrain structure. It is the visual processing center.
Distal	Students can set both long-term (distal) and short-term (proximal) goals .
Positron emission tomography	Positron Emission Tomography measures emissions from radioactively labeled chemicals that have been injected into the bloodstream. The greatest benefit is that different compounds can show blood flow and oxygen and glucose metabolism in the tissues of the working brain.
Cerebral cortex	The cerebral cortex is the outermost layer of the cerebrum and has a grey color. It is made up of four lobes and it is involved in many complex brain functions including memory, perceptual awareness, "thinking", language and consciousness. The cerebral cortex receives sensory information from many different sensory organs eg: eyes, ears, etc. and processes the information.
Hypothesis	A specific statement about behavior or mental processes that is testable through research is

	a hypothesis.
Functional magnetic resonance imaging	Functional Magnetic Resonance Imaging describes the use of MRI to measure hemodynamic signals related to neural activity in the brain or spinal cord of humans or other animals. It is one of the most recently developed forms of brain imaging.
Insight	Insight refers to a sudden awareness of the relationships among various elements that had previously appeared to be independent of one another.
Species	Species refers to a reproductively isolated breeding population.
Seizure	A seizure is a temporary alteration in brain function expressed as a changed mental state, tonic or clonic movements and various other symptoms. They are due to temporary abnormal electrical activity of a group of brain cells.
Control subjects	Control subjects are participants in an experiment who do not receive the treatment effect but for whom all other conditions are held comparable to those of experimental subjects.
Corpus callosum	The corpus callosum is the largest white matter structure in the brain. It consists of mostly of contralateral axon projections. The corpus callosum connects the left and right cerebral hemispheres. Most communication between regions in different halves of the brain are carried over the corpus callosum.
Tactile	Pertaining to the sense of touch is referred to as tactile.
Nerve	A nerve is an enclosed, cable-like bundle of nerve fibers or axons, which includes the glia that ensheath the axons in myelin. Neurons are sometimes called nerve cells, though this term is technically imprecise since many neurons do not form nerves.
Split-brain	Split-brain is a lay term to describe the result when the corpus callosum connecting the two halves of the brain is severed to some degree. The surgical operation to produce this condition is called corpus callosotomy. It is rarely performed, usually to treat epilepsy; people with generalized seizures who injure themselves during falls may have less violent seizures, and hence less injury during those seizures, after the procedure.
Gazzaniga	Gazzaniga worked under the guidance of Roger Sperry, with primary responsibility for initiating human split-brain research. In his subsequent work he has made important advances in our understanding of functional lateralization in the brain and how the cerebral hemispheres communicate with one another.
Theories	Theories are logically self-consistent models or frameworks describing the behavior of a certain natural or social phenomenon. They are broad explanations and predictions concerning phenomena of interest.
Chronic	Chronic refers to a relatively long duration, usually more than a few months.
Dorsal stream	The dorsal stream is a pathway for visual information which flows through the visual cortex, the part of the brain which provides visual processing. It is involved in spatial awareness: recognizing where objects are in space.
Elaboration	The extensiveness of processing at any given level of memory is called elaboration. The use of elaboration changes developmentally. Adolescents are more likely to use elaboration spontaneously than children.
Inference	Inference is the act or process of drawing a conclusion based solely on what one already knows.
Neurochemistry	Neurochemistry is a branch of neuroscience that is heavily devoted to the study of neurochemicals. A neurochemical is an organic molecule that participates in neural activity. This term is often used to refer to neurotransmitters and other molecules such as neuro-

	active drugs that influence neuron function.
Learning	Learning is a relatively permanent change in behavior that results from experience. Thus, to attribute a behavioral change to learning, the change must be relatively permanent and must result from experience.
Hormone	A hormone is a chemical messenger from one cell (or group of cells) to another. The best known are those produced by endocrine glands, but they are produced by nearly every organ system. The function of hormones is to serve as a signal to the target cells; the action of the hormone is determined by the pattern of secretion and the signal transduction of the receiving tissue.
Longitudinal study	Longitudinal study is a type of developmental study in which the same group of participants is followed and measured for an extended period of time, often years.
Menarche	Menarche is the first menstrual period as a girl's body progresses through the changes of puberty. Menarche usually occurs about two years after the first changes of breast development.
Maturation	The orderly unfolding of traits, as regulated by the genetic code is called maturation.
Cognitive skills	Cognitive skills such as reasoning, attention, and memory can be advanced and sustained through practice and training.
Puberty	Puberty refers to the process of physical changes by which a child's body becomes an adult body capable of reproduction.
Gene	A gene is an ultramicroscopic area of the chromosome. It is the smallest physical unit of the DNA molecule that carries a piece of hereditary information.
Y chromosome	The Y chromosome is one of the two sex chromosomes in humans and most other mammals. The sex chromosomes are one of the 23 pairs of human chromosomes. The Y chromosome contains the fewest genes of any of the chromosomes. It contains the genes that cause testis development, thus determining maleness. It is usually contributed by the father.
Androgen	Androgen is the generic term for any natural or synthetic compound, usually a steroid hormone, that stimulates or controls the development and maintenance of masculine characteristics in vertebrates by binding to androgen receptors.
Testes	Testes are the male reproductive glands or gonads; this is where sperm develop and are stored.
Natural selection	Natural selection is a process by which biological populations are altered over time, as a result of the propagation of heritable traits that affect the capacity of individual organisms to survive and reproduce.
Aphasia	Aphasia is a loss or impairment of the ability to produce or comprehend language, due to brain damage. It is usually a result of damage to the language centers of the brain.
Population	Population refers to all members of a well-defined group of organisms, events, or things.
Anesthesia	Anesthesia is the process of blocking the perception of pain and other sensations. This allows patients to undergo surgery and other procedures without the distress and pain they would otherwise experience.
Epilepsy	Epilepsy is a chronic neurological condition characterized by recurrent unprovoked neural discharges. It is commonly controlled with medication, although surgical methods are used as well.
Barbiturate	A barbiturate is a drug that acts as a central nervous system (CNS) depressant, and by virtue of this produces a wide spectrum of effects, from mild sedation to anesthesia.

Sensation	Sensation is the first stage in the chain of biochemical and neurologic events that begins with the impinging of a stimulus upon the receptor cells of a sensory organ, which then leads to perception, the mental state that is reflected in statements like "I see a uniformly blue wall."
Myelin	Myelin is an electrically insulating fatty layer that surrounds the axons of many neurons, especially those in the peripheral nervous system. The main consequence of a myelin sheath is an increase in the speed at which impulses propagate along the myelinated fiber. The sheath continues to develop throughout childhood.
Simulation	A simulation is an imitation of some real device or state of affairs. Simulation attempts to represent certain features of the behavior of a physical or abstract system by the behavior of another system.
Toddler	A toddler is a child between the ages of one and three years old. During this period, the child learns a great deal about social roles and develops motor skills; to toddle is to walk unsteadily.
Neurologist	A physician who studies the nervous system, especially its structure, functions, and abnormalities is referred to as neurologist.
Charles Spearman	Charles Spearman is known for his work in statistics, as a pioneer of factor analysis, and for his rank correlation coefficient. He also did seminal work on models for human intelligence, including discovering that disparate cognitive test scores reflect a single general factor and coining the term g factor.
G factor	Spearman's term for a general intellectual ability that underlies all mental operations to some degree is called the g factor.
Neuropsychologist	A psychologist concerned with the relationships among cognition, affect, behavior, and brain function is a neuropsychologist.
Howard Gardner	Howard Gardner is best known for his theory of multiple intelligences. He outlined seven dimensions of intelligence: Visual / Spatial Intelligence, Musical Intelligence, Verbal Intelligence, Logical/Mathematical Intelligence, Interpersonal Intelligence, Intrapersonal Intelligence, and Bodily / Kinesthetic Intelligence. Since the publication of Frames of Mind, he has additionally identified the 8th and 9th dimensions of intelligence: Naturalist Intelligence and Existentialist Intelligence.
Intrapersonal	Intrapersonal is the ability to recognize, define, and pursue inner feelings and thoughts, as in poetry and self-knowledge.
Intelligence test	An intelligence test is a standardized means of assessing a person's current mental ability, for example, the Stanford-Binet test and the Wechsler Adult Intelligence Scale.
Convergent thinking	Convergent thinking is a thought process that attempts to gradually narrow in on the single best solution to a problem.
J. P. Guilford	J. P. Guilford rejected Charles Spearman's view that intelligence could be characterised in a single numerical parameter and proposed that three dimensions were necessary for accurate description:
Divergent thinking	A thought process that attempts to generate multiple solutions to problems is called divergent thinking.
IQ test	An IQ test is a standardized test developed to measure a person's cognitive abilities ("intelligence") in relation to their age group.
Heredity	Heredity is the transfer of characteristics from parent to offspring through their genes.
Synapse	A synapse is specialized junction through which cells of the nervous system signal to one

	another and to non-neuronal cells such as muscles or glands.
Postnatal	Postnatal is the period beginning immediately after the birth of a child and extending for about six weeks. The period is also known as postpartum and, less commonly, puerperium.
Quantitative	A quantitative property is one that exists in a range of magnitudes, and can therefore be measured. Measurements of any particular quantitative property are expressed as as a specific quantity, referred to as a unit, multiplied by a number.
Glial	Glial cells are non-neuronal cells that provide support and nutrition, maintain homeostasis, form myelin, and participate in signal transmission in the nervous system.
Theory of multiple intelligences	Gardner's theory of multiple intelligences suggests several different kinds of "intelligence" exist in humans, each relating to a different sphere of human life and activity. .
Guilford	Guilford observed that most individuals display a preference for either convergent or divergent thinking. Scientists and engineers typically prefer the former and artists and performers, the latter.
Consciousness	The awareness of the sensations, thoughts, and feelings being experienced at a given moment is called consciousness.
Subjective experience	Subjective experience refers to reality as it is perceived and interpreted, not as it exists objectively.
Reflection	Reflection is the process of rephrasing or repeating thoughts and feelings expressed, making the person more aware of what they are saying or thinking.
Variability	Statistically, variability refers to how much the scores in a distribution spread out, away from the mean.
Trait	An enduring personality characteristic that tends to lead to certain behaviors is called a trait. The term trait also means a genetically inherited feature of an organism.
Dissociation	Dissociation is a psychological state or condition in which certain thoughts, emotions, sensations, or memories are separated from the rest.
Rods	Rods are cylindrical shaped photoreceptors that are sensitive to the intensity of light. Rods require less light to function than cone cells, and therefore are the primary source of visual information at night.
Cognitive science	Cognitive Science is the scientific study of the mind and brain and how they give rise to behavior. The field is highly interdisciplinary and is closely related to several other areas, including psychology, artificial intelligence, linguistics and psycholinguistics, philosophy, neuroscience, logic, robotics, anthropology and biology.
Forebrain	The forebrain is the highest level of the brain. Key structures in the forebrain are the limbic system, thalamus, basal ganglia, hypothalamus, and cerebral cortex.
Agnosia	Agnosia is a loss of ability to recognize objects, persons, sounds, shapes or smells while the specific sense is not defective nor is there any significant memory loss. It is usually associated with brain injury or neurological illness, particularly after damage to the temporal lobe.
Self-awareness	Realization that one's existence and functioning are separate from those of other people and things is called self-awareness.
Chalmers	Chalmers makes the distinction between easy problems of consciousness and the hard problem, which could be stated "why does awareness of sensory information exist at all?" He argues against any identification of the experience of awareness with purely physical processes.

Tact	The word tact, another of Skinner's intentionally "nonsense" words, comes from the notion of the child's making "conTACT" with the nonverbal environment. The tact is verbal behavior that is under the control of the nonverbal environment and includes nouns, actions, adjectives, pronouns, relations, and others.
Syndrome	The term syndrome is the association of several clinically recognizable features, signs, symptoms, phenomena or characteristics which often occur together, so that the presence of one feature indicates the presence of the others.
Amnesia	Amnesia is a condition in which memory is disturbed. The causes of amnesia are organic or functional. Organic causes include damage to the brain, through trauma or disease, or use of certain (generally sedative) drugs.
Verbal Behavior	Verbal Behavior is a book written by B.F. Skinner in which the author presents his ideas on language. For Skinner, speech, along with other forms of communication, was simply a behavior. Skinner argued that each act of speech is an inevitable consequence of the speaker's current environment and his behavioral and sensory history.
Central nervous system	The vertebrate central nervous system consists of the brain and spinal cord.
Anatomy	Anatomy is the branch of biology that deals with the structure and organization of living things. It can be divided into animal anatomy (zootomy) and plant anatomy (phytonomy). Major branches of anatomy include comparative anatomy, histology, and human anatomy.
Metabolism	Metabolism is the biochemical modification of chemical compounds in living organisms and cells.
Penfield	Penfield treated patients with severe epilepsy by destroying nerve cells in the brain. Before operating, he stimulated the brain with electrical probes while the patients were conscious on the operating table, and observed their responses. It allowed him to create maps of sensory and motor functions.
Nervous system	The body's electrochemical communication circuitry, made up of billions of neurons is a nervous system.

Brain	The brain controls and coordinates most movement, behavior and homeostatic body functions such as heartbeat, blood pressure, fluid balance and body temperature. Functions of the brain are responsible for cognition, emotion, memory, motor learning and other sorts of learning. The brain is primarily made up of two types of cells: glia and neurons.
Nervous system	The body's electrochemical communication circuitry, made up of billions of neurons is a nervous system.
Learning	Learning is a relatively permanent change in behavior that results from experience. Thus, to attribute a behavioral change to learning, the change must be relatively permanent and must result from experience.
Perception	Perception is the process of acquiring, interpreting, selecting, and organizing sensory information.
Species	Species refers to a reproductively isolated breeding population.
Neuron	The neuron is the primary cell of the nervous system. They are found in the brain, the spinal cord, in the nerves and ganglia of the peripheral nervous system. It is a specialized cell that conducts impulses through the nervous system and contains three major parts: cell body, dendrites, and an axon. It can have many dendrites but only one axon.
Evolution	Commonly used to refer to gradual change, evolution is the change in the frequency of alleles within a population from one generation to the next. This change may be caused by different mechanisms, including natural selection, genetic drift, or changes in population (gene flow).
Dendrite	A dendrite is a slender, typically branched projection of a nerve cell, or "neuron," which conducts the electrical stimulation received from other cells to the body or soma of the cell from which it projects. This stimulation arrives through synapses, which typically are located near the tips of the dendrites and away from the soma.
Axon	An axon, or "nerve fiber," is a long slender projection of a nerve cell, or "neuron," which conducts electrical impulses away from the neuron's cell body or soma.
Axon terminal	A swelling at the end of an axon that is designed to release a chemical substance onto another neuron, muscle cell, or gland cell is called the axon terminal.
Synapse	A synapse is specialized junction through which cells of the nervous system signal to one another and to non-neuronal cells such as muscles or glands.
Synaptic cleft	Synaptic cleft refers to a microscopic gap between the terminal button of a neuron and the cell membrane of another neuron.
Receptor	A sensory receptor is a structure that recognizes a stimulus in the internal or external environment of an organism. In response to stimuli the sensory receptor initiates sensory transduction by creating graded potentials or action potentials in the same cell or in an adjacent one.
Receptor site	A location on the dendrite of a receiving neuron that is tailored to receive a specific neurotransmitter is a receptor site.
Nerve	A nerve is an enclosed, cable-like bundle of nerve fibers or axons, which includes the glia that ensheath the axons in myelin. Neurons are sometimes called nerve cells, though this term is technically imprecise since many neurons do not form nerves.
Drug addiction	Drug addiction, or substance dependence is the compulsive use of drugs, to the point where the user has no effective choice but to continue use.
Hormone	A hormone is a chemical messenger from one cell (or group of cells) to another. The best known are those produced by endocrine glands, but they are produced by nearly every organ system. The function of hormones is to serve as a signal to the target cells; the action of

	the hormone is determined by the pattern of secretion and the signal transduction of the receiving tissue.
Stages	Stages represent relatively discrete periods of time in which functioning is qualitatively different from functioning at other periods.
Gene	A gene is an ultramicroscopic area of the chromosome. It is the smallest physical unit of the DNA molecule that carries a piece of hereditary information.
Genetics	Genetics is the science of genes, heredity, and the variation of organisms.
Dorsal stream	The dorsal stream is a pathway for visual information which flows through the visual cortex, the part of the brain which provides visual processing. It is involved in spatial awareness: recognizing where objects are in space.
Motivation	In psychology, motivation is the driving force (desire) behind all actions of an organism.
Construct	A generalized concept, such as anxiety or gravity, is a construct.
Basal ganglia	The basal ganglia are a group of nuclei in the brain associated with motor and learning functions.
Hippocampus	The hippocampus is a part of the brain located inside the temporal lobe. It forms a part of the limbic system and plays a part in memory and navigation.
Amygdala	Located in the brain's medial temporal lobe, the almond-shaped amygdala is believed to play a key role in the emotions. It forms part of the limbic system and is linked to both fear responses and pleasure. Its size is positively correlated with aggressive behavior across species.
Emotion	An emotion is a mental states that arise spontaneously, rather than through conscious effort. They are often accompanied by physiological changes.
Central nervous system	The vertebrate central nervous system consists of the brain and spinal cord.
Anatomy	Anatomy is the branch of biology that deals with the structure and organization of living things. It can be divided into animal anatomy (zootomy) and plant anatomy (phytonomy). Major branches of anatomy include comparative anatomy, histology, and human anatomy.
Neural network	A clusters of neurons that is interconnected to process information is referred to as a neural network.
Cerebral hemisphere	Either of the two halves that make up the cerebrum is referred to as a cerebral hemisphere. The hemispheres operate together, linked by the corpus callosum, a very large bundle of nerve fibers, and also by other smaller commissures.
Right hemisphere	The brain is divided into left and right cerebral hemispheres. The right hemisphere of the cortex controls the left side of the body.
Parietal lobe	The parietal lobe is positioned above (superior to) the occipital lobe and behind (posterior to) the frontal lobe. It plays important roles in integrating sensory information from various senses, and in the manipulation of objects.
Temporal lobe	The temporal lobe is part of the cerebrum. It lies at the side of the brain, beneath the lateral or Sylvian fissure. Adjacent areas in the superior, posterior and lateral parts of the temporal lobe are involved in high-level auditory processing.
Hyperpolarization	Hyperpolarization is any change in a cell's membrane potential that makes it more polarized. That is, hyperpolarization is an increase in the absolute value of a cell's membrane potential.

Depolarization	Depolarization is any change that decreases the absolute value of a cell's membrane potential. Thus, changes in membrane voltage in which the membrane potential becomes less positive or less negative are both depolarizations.
Depression	In everyday language depression refers to any downturn in mood, which may be relatively transitory and perhaps due to something trivial. This is differentiated from Clinical depression which is marked by symptoms that last two weeks or more and are so severe that they interfere with daily living.
Chorea	Chorea is the occurrence of continuous rapid, jerky, involuntary movements that may involve the face and limb and result in an inability to maintain a posture. It is also known as St. Vitus Dance disease and is seen mostly in children.
Hallucination	A hallucination is a sensory perception experienced in the absence of an external stimulus, as distinct from an illusion, which is a misperception of an external stimulus. They may occur in any sensory modality - visual, auditory, olfactory, gustatory, tactile, or mixed.
Tremor	Tremor is the rhythmic, oscillating shaking movement of the whole body or just a certain part of it, caused by problems of the neurons responsible from muscle action.
Somatosensory cortex	The primary somatosensory cortex is across the central sulcus and behind the primary motor cortex configured to generally correspond with the arrangement of nearby motor cells related to specific body parts. It is the main sensory receptive area for the sense of touch.
Somatosensory	Somatosensory system consists of the various sensory receptors that trigger the experiences labelled as touch or pressure, temperature, pain, and the sensations of muscle movement and joint position including posture, movement, and facial expression.
Affect	A subjective feeling or emotional tone often accompanied by bodily expressions noticeable to others is called affect.
Hypothalamus	The hypothalamus is a region of the brain located below the thalamus, forming the major portion of the ventral region of the diencephalon and functioning to regulate certain metabolic processes and other autonomic activities.
Adaptive behavior	An adaptive behavior increases the probability of the individual or organism to survive or exist within its environment.
Frontal lobe	The frontal lobe comprises four major folds of cortical tissue: the precentral gyrus, superior gyrus and the middle gyrus of the frontal gyri, the inferior frontal gyrus. It has been found to play a part in impulse control, judgement, language, memory, motor function, problem solving, sexual behavior, socialization and spontaneity.
Forebrain	The forebrain is the highest level of the brain. Key structures in the forebrain are the limbic system, thalamus, basal ganglia, hypothalamus, and cerebral cortex.
Anxiety disorder	Anxiety disorder is a blanket term covering several different forms of abnormal anxiety, fear, phobia and nervous condition, that come on suddenly and prevent pursuing normal daily routines.
Schizophrenia	Schizophrenia is characterized by persistent defects in the perception or expression of reality. A person suffering from untreated schizophrenia typically demonstrates grossly disorganized thinking, and may also experience delusions or auditory hallucinations
Brain-imaging	Brain-imaging is a fairly recent discipline within medicine and neuroscience. Brain imaging falls into two broad categories -- structural imaging and functional imaging.
Infancy	The developmental period that extends from birth to 18 or 24 months is called infancy.
Neuroanatomy	Neuroanatomy is the study of the anatomy of the central nervous system.

Neuroscience	A field that combines the work of psychologists, biologists, biochemists, medical researchers, and others in the study of the structure and function of the nervous system is neuroscience.
Cajal	Cajal, Nobel laureate, is considered to be the father of modern neuroscience. He postulated that the nervous system is made up of billions of separate nerve cells (neurons) and that nerve cells are polarized. He suggested that neurons communicate with each other via specialized junctions. This hypothesis became the basis of the neuron doctrine.
Abnormal behavior	An action, thought, or feeling that is harmful to the person or to others is called abnormal behavior.
Personality	Personality refers to the pattern of enduring characteristics that differentiates a person, the patterns of behaviors that make each individual unique.
Homosexual	Homosexual refers to a sexual orientation characterized by aesthetic attraction, romantic love, and sexual desire exclusively for members of the same sex or gender identity.
Population	Population refers to all members of a well-defined group of organisms, events, or things.
Chronic disorders	Chronic disorders are characterized by slow onset and long duration. They are rare in early adulthood, they increase during middle adulthood, and they become common in late adulthood.
Cerebral palsy	Cerebral palsy is a group of permanent disorders associated with developmental brain injuries that occur during fetal development, birth, or shortly after birth. It is characterized by a disruption of motor skills, with symptoms such as spasticity, paralysis, or seizures.
Spinal cord	The spinal cord is a part of the vertebrate nervous system that is enclosed in and protected by the vertebral column (it passes through the spinal canal). It consists of nerve cells. The spinal cord carries sensory signals and motor innervation to most of the skeletal muscles in the body.
Epilepsy	Epilepsy is a chronic neurological condition characterized by recurrent unprovoked neural discharges. It is commonly controlled with medication, although surgical methods are used as well.
Insomnia	Insomnia is a sleep disorder characterized by an inability to sleep and/or to remain asleep for a reasonable period during the night.
Chronic	Chronic refers to a relatively long duration, usually more than a few months.
Stroke	A stroke occurs when the blood supply to a part of the brain is suddenly interrupted by occlusion, by hemorrhage, or other causes
Trauma	Trauma refers to a severe physical injury or wound to the body caused by an external force, or a psychological shock having a lasting effect on mental life.
Acute	Acute means sudden, sharp, and abrupt. Usually short in duration.
Dopamine	Dopamine is critical to the way the brain controls our movements and is a crucial part of the basal ganglia motor loop. It is commonly associated with the 'pleasure system' of the brain, providing feelings of enjoyment and reinforcement to motivate us to do, or continue doing, certain activities.
Metabolism	Metabolism is the biochemical modification of chemical compounds in living organisms and cells.
Glutamate	Glutamate is one of the 20 standard amino acids used by all organisms in their proteins. It is critical for proper cell function, but it is not an essential nutrient in humans because it can be manufactured from other compounds.
Affective	Affective is the way people react emotionally, their ability to feel another living thing's

pain or joy.

Clinical depression — Although nearly any mood with some element of sadness may colloquially be termed a depression, clinical depression is more than just a temporary state of sadness. Symptoms lasting two weeks or longer in duration, and of a severity that they begin to interfere with daily living.

Suicide — Suicide behavior is rare in childhood but escalates in adolescence. The suicide rate increases in a linear fashion from adolescence through late adulthood.

Guilt — Guilt describes many concepts related to a negative emotion or condition caused by actions which are believed to be, morally wrong. According to Freud, the avoidance of guilt is the basis for moral behavior.

Habit — A habit is a response that has become completely separated from its eliciting stimulus. Early learning theorists used the term to describe S-R associations, however not all S-R associations become a habit, rather many are extinguished after reinforcement is withdrawn.

Mania — Mania is a medical condition characterized by severely elevated mood. Mania is most usually associated with bipolar disorder, where episodes of mania may cyclically alternate with episodes of depression.

Bipolar disorder — Bipolar Disorder is a mood disorder typically characterized by fluctuations between manic and depressive states; and, more generally, atypical mood regulation and mood instability.

Norepinephrine — Norepinephrine is released from the adrenal glands as a hormone into the blood, but it is also a neurotransmitter in the nervous system. As a stress hormone, it affects parts of the human brain where attention and impulsivity are controlled. Along with epinephrine, this compound effects the fight-or-flight response, activating the sympathetic nervous system to directly increase heart rate, release energy from fat, and increase muscle readiness.

Serotonin — Serotonin, a neurotransmitter, is believed to play an important part of the biochemistry of depression, bipolar disorder and anxiety. It is also believed to be influential on sexuality and appetite.

Theories — Theories are logically self-consistent models or frameworks describing the behavior of a certain natural or social phenomenon. They are broad explanations and predictions concerning phenomena of interest.

Antidepressant — An antidepressant is a medication used primarily in the treatment of clinical depression. They are not thought to produce tolerance, although sudden withdrawal may produce adverse effects. They create little if any immediate change in mood and require between several days and several weeks to take effect.

Transduction — Transduction in the nervous system typically refers to synaptic events wherein an electrical signal, known as an action potential, is converted into a chemical one via the release of neurotransmitters. Conversely, in sensory transduction a chemical or physical stimulus is transduced by sensory receptors into an electrical signal.

Tumor — A tumor is an abnormal growth that when located in the brain can either be malignant and directly destroy brain tissue, or be benign and disrupt functioning by increasing intracranial pressure.

Electrode — Any device used to electrically stimulate nerve tissue or to record its activity is an electrode.

Thalamus — An area near the center of the brain involved in the relay of sensory information to the cortex and in the functions of sleep and attention is the thalamus.

Cell — Cell differentiation is a concept from developmental biology describing the process by which

differentiation	cells acquire a "type". The morphology of a cell may change dramatically during differentiation, but the genetic material remains the same, with few exceptions.
Stem cells	Stem cells are primal undifferentiated cells which retain the ability to differentiate into other cell types. This ability allows them to act as a repair system for the body, replenishing other cells as long as the organism is alive.
Multipotent stem cells	Multipotent stem cells can give rise to several other cell types, but those types are limited in number.
Phenotype	The phenotype of an individual organism is either its total physical appearance and constitution, or a specific manifestation of a trait, such as size or eye color, that varies between individuals. Phenotype is determined to some extent by genotype, or by the identity of the alleles that an individual carries at one or more positions on the chromosomes.
Chemical imbalance	Chemical imbalance refers to relative excess or deficit in brain chemicals, such as neurotransmitters, that may be implicated in some psychological disorders.
Tardive dyskinesia	Tardive dyskinesia is a serious neurological disorder caused by the long-term use of traditional antipsychotic drugs.
Neuroleptics	Neuroleptics are a group of drugs used to treat psychosis. Common conditions with which neuroleptics might be used include schizophrenia, mania and delusional disorder. They also have some effects as mood stabilizers, leading to their frequent use in treating mood disorder (particularly bipolar disorder) even when no signs of psychosis are present.
Relearning	Relearning refers to a measure of retention used in experiments on memory. Material is usually relearned more quickly than it is learned initially.
Sigmund Freud	Sigmund Freud was the founder of the psychoanalytic school, based on his theory that unconscious motives control much behavior, that particular kinds of unconscious thoughts and memories are the source of neurosis, and that neurosis could be treated through bringing these unconscious thoughts and memories to consciousness in psychoanalytic treatment.
Insight	Insight refers to a sudden awareness of the relationships among various elements that had previously appeared to be independent of one another.
Behavioral therapy	The treatment of a mental disorder through the application of basic principles of conditioning and learning is called behavioral therapy.
Mental illness	Mental illness is the term formerly used to mean psychological disorder but less preferred because it implies that the causes of the disorder can be found in a medical disease process.
Mental disorder	Mental disorder refers to a disturbance in a person's emotions, drives, thought processes, or behavior that involves serious and relatively prolonged distress and/or impairment in ability to function, is not simply a normal response to some event or set of events in the person's environment.
Lesion	A lesion is a non-specific term referring to abnormal tissue in the body. It can be caused by any disease process including trauma (physical, chemical, electrical), infection, neoplasm, metabolic and autoimmune.
Severe mental retardation	A limitation in mental development as measured on the Wechsler Adult Intelligence Scale with scores between 20 -34 is called severe mental retardation.
Phenylalanine	Phenylalanine is an essential amino acid. The genetic disorder phenylketonuria is an inability to metabolize phenylalanine.
Tyrosine	Tyrosine is one of the 20 amino acids that are used by cells to synthesize proteins. It plays a key role in signal transduction, since it can be tagged (phosphorylated) with a phosphate group by protein kinases to alter the functionality and activity of certain enzymes.

Mental retardation	Mental retardation refers to having significantly below-average intellectual functioning and limitations in at least two areas of adaptive functioning. Many categorize retardation as mild, moderate, severe, or profound.
Neurologist	A physician who studies the nervous system, especially its structure, functions, and abnormalities is referred to as neurologist.
Phenylketonuria	Phenylketonuria is a genetic disorder in which an individual cannot properly metabolize amino acids. The disorder is now easily detected but, if left untreated, results in mental retardation and hyperactivity.
Amino acid	Amino acid is the basic structural building unit of proteins. They form short polymer chains called peptides or polypeptides which in turn form structures called proteins.
Enzyme	An enzyme is a protein that catalyzes, or speeds up, a chemical reaction. Enzymes are essential to sustain life because most chemical reactions in biological cells would occur too slowly, or would lead to different products, without enzymes.
Behavioral observation	A form of behavioral assessment that entails careful observation of a person's overt behavior in a particular situation is behavioral observation.
Agonist	Agonist refers to a drug that mimics or increases a neurotransmitter's effects.
Plasticity	The capacity for modification and change is referred to as plasticity.
Dendritic spine	A dendritic spine is a small membranous extrusion that protrudes from a dendrite and forms one half of a synapse. Changes in dendritic spine density underlie many brain functions, including motivation, learning, and memory. In particular, long-term memory is mediated in part by the growth of new dendritic spines to reinforce a particular neural pathway.
Pathology	Pathology is the study of the processes underlying disease and other forms of illness, harmful abnormality, or dysfunction.
Animal model	An animal model usually refers to a non-human animal with a disease that is similar to a human condition.
Tay-Sachs disease	Tay-Sachs disease is a fatal genetic disorder, inherited in an autosomal recessive pattern, in which harmful quantities of a fatty substance called ganglioside GM2 accumulate in the nerve cells in the brain.
Predisposition	Predisposition refers to an inclination or diathesis to respond in a certain way, either inborn or acquired. In abnormal psychology, it is a factor that lowers the ability to withstand stress and inclines the individual toward pathology.
Catatonic behavior	Catatonic behavior is a symptom of schizophrenia in which the person is unresponsive to the environment. It may take the form of active resistance, excited motor activity, or a complete lack of movement or awareness of the environment.
Delusion	A false belief, not generally shared by others, and that cannot be changed despite strong evidence to the contrary is a delusion.
Antipsychotic	The term antipsychotic is applied to a group of drugs used to treat psychosis.
Prognosis	A forecast about the probable course of an illess is referred to as prognosis.
Prefrontal cortex	The prefrontal cortex is the anterior part of the frontal lobes of the brain, lying in front of the motor and associative areas. It has been implicated in planning complex cognitive behaviors, personality expression and moderating correct social behavior. The prefrontal cortex continues to develop until around age 6.
Cingulate cortex	The part of the limbic system that is believed to process cognitive information in emotion is the cingulate cortex. The cingulate cortex is part of the brain and situated roughly in the

middle of the cortex.

Neurochemistry

Neurochemistry is a branch of neuroscience that is heavily devoted to the study of neurochemicals. A neurochemical is an organic molecule that participates in neural activity. This term is often used to refer to neurotransmitters and other molecules such as neuro-active drugs that influence neuron function.

Lightning Source UK Ltd.
Milton Keynes UK
UKHW030009151020
371596UK00002B/18

9 781428 801592